MATH WORKBOOK FOR FOODSERVICE/LODGING

THIRD EDITION

M. C. McDOWELL
H. W. CRAWFORD

JOHN WILEY & SONS, INC.
New York Chichester Weinheim Brisbane Singapore Toronto

This text is printed on acid-free paper. ⊗

Copyright © 1988, 1981, 1971 by John Wiley & Sons, Inc. All rights reserved.

Published simultaneously in Canada.

No part of this publication may be reproduced, stored in a retrieval system or transmitted in any form or by any means, electronic, mechanical, photocopying, recording, scanning or otherwise, except as permitted under Sections 107 or 108 of the 1976 United States Copyright Act, without either the prior written permission of the Publisher, or authorization through payment of the appropriate per-copy fee to the Copyright Clearance Center, 222 Rosewood Drive, Danvers, MA 01923, (978) 750-8400, fax (978) 750-4744. Requests to the Publisher for permission should be addressed to the Permissions Department, John Wiley & Sons, Inc., 605 Third Avenue, New York, NY 10158-0012, (212) 850-6011, fax (212) 850-6008, E-Mail: PERMREQ@WILEY.COM.

This publication is designed to provide accurate and authoritative information in regard to the subject matter covered. It is sold with the understanding that the publisher is not engaged in rendering legal, accounting, or other professional services. If legal advice or other expert assistance is required, the services of a competent professional person should be sought.

16 15 14 13 12 11 10

Library of Congress Cataloging-in-Publication Data:

McDowell, M. C. (Milton C.) 1920-
  Math workbook for foodservice/lodging.

  1. Business mathematics—Food service—Problems, exercises, etc. 2. Business mathematics —Hotels, taverns, etc.—Problems, exercises, etc. 3. Business mathematics—Motels— Problems, exercises, etc. I. Crawford, H. W. (Hollie W.), 1919- II. Title.
HF5695.5.F616M317 1987   513'.93'024642   87-15960
ISBN 0-471-28875-6

CONTENTS

FOR THE STUDENT — 1

Introduction	3
Fit Yourself into the Future	5
Contract	6
Attendance Record	7
Weekly Test Record	8
List of Symbols	9
Addition Table	11
Multiplication Table I	12
Multiplication Table II	13
Subtraction Table	14
Equivalent Fractions	15
Prime Numbers	16

ESSENTIALS OF MATHEMATICS — 17

Diagnostic Tests	19
Number and Numeral	23
How to Add	27
How to Multiply	32
How to Subtract	41
How to Make Change	45
How to Divide	47
How to Work with Fractions	52
How to Operate with Decimals	60
How to Operate with Denominate Numbers	73
How-to of Ratio, Proportion, and Percent	76
Exercises	82
Weekly Tests I–XL	98

BUSINESS SITUATIONS INVOLVING MATHEMATICS — 109

Sanitation	111
Wage Scales	116
Inflation and the Consumer Price Index	119
Labor Costs	125
Working Conditions in the Industry	128
Using Standards in Foodservice	133
Converting Standard Recipes	135
As Purchased Compared with Edible Portion	138
Costing Standard Recipes	143
Cook's Daily Reports	147
Break-Even Analysis (To Find Break-Even Point in Sales)	154

Seating Turnover	157
Pricing the Menu (Mark-Up)	160
What Is an Ounce?	163
Wine with Your Meal?	164
Drill for Baker's Scale	167
Guest Checks	169
Cash Receipts Report	179
Business Forms	181
Storeroom Fundamentals	190
Financial Statements	192
Loan Payments and Interest	202
Income Taxes—"Pay as you Go"	204
Local Taxes—How Business Helps Carry the Load	215
Types of Business Ownership	219
Plans for an Inn—A Scale Drawing	223
What Employers Look for When Hiring	225
Activity Problems	227

SUPPLEMENTARY TABLES AND CHARTS 229

Tables of Measure	231
Egg Specifications	232
Fractional Equivalents for Use in Converting Recipes	234
Container Portion and Conversion Chart	235
Conversion Tables	236
Decimal Equivalents of Common Fractions	238
Ounces and Eighths of Ounces Expressed as Decimal Parts of Pound	239
Table of Approximate Weights and Measures of Common Foods	240
Days Expressed as Decimal Part of Month	242
Abbreviations and Equivalents	243
Table of Weights and their Approximate Equivalents in Measure	244
Size Portions and Serving Utensils Required	247
Portion Control Data	251
Common Container Sizes	256
Weight Conversion Charts	261

INDEX 263

Milton C. McDowell holds a B.S. in Business Administration from the University of San Francisco and an M.A. in Secondary Education from Stanford University. His business experience includes a number of years as a licensed public accountant. During a thirty-year teaching career, Mr. McDowell taught in public and private high schools and adult schools. In 1965, while business department head at Capuchino High School in San Bruno, California, he and three other faculty members designed an interdisciplinary program for food education and service training (FEAST), which is still being taught. He has also been a member of the Project Feast staff at City College of San Francisco, where he served as a consultant in course planning for teachers attending Project Feast workshops. Mr. McDowell is currently administrator for the Hotel and Restaurant Foundation in San Francisco.

FOR THE STUDENT

INTRODUCTION

This workbook is designed to help you learn to solve problems and to do a lot of learning on your own. Read this section carefully. In it you will find a Contract which permits you to negotiate for a grade with an instructor, an Attendance Record you can keep to see if you would hire yourself for a job, a Weekly Test Record for keeping your own record of grades. These records are followed in the next section by several useful tables. Then there are four Diagnostic Tests to be taken, one at a time.

First, take the diagnostic test on Whole Numbers. If you discover that you need help, turn to the "How to" chapters and learn how to do the problems missed, then go to the Exercises chapter. These exercises have detailed breakdowns in each operation for mathematical problem solving. You can choose the exercises yourself, or the instructor can assign them to you. Progress through the exercises whenever you discover an area of need from the Diagnostic Tests. When you have mastered whole numbers, move to the diagnostic test on fractions, then decimals, and finally percentage. You may take the Diagnostic Tests more than once.

Choose a certain day each week to take a Weekly Test, recording your score on the record sheet provided. When you are ready, move into the units covering the foodservice/lodging industry that begin immediately following Weekly Test XL.

Also, become familiar with all the supplementary tables and charts at the end of this book. They can be most useful tools in your work.

Finally, it is in your best interests to keep the records here accurately, to find your own problem areas, and to evaluate your work as honestly as you can. The authors recommend that you make every effort to read and refer to as many books as possible. If you have a reading problem, do all you can to increase your ability to read. It may seem strange to discuss reading in a math workbook but—

MATHEMATICS AND READING ARE KEY SUBJECTS FOR MAKING YOU EMPLOYABLE

To get and hold a job in the foodservice/lodging industry, your knowledge of mathematics should include proficiency in the fundamental skills of adding, subtracting, multiplying, and dividing of whole and rational numbers. In addition, you will need some knowledge in the areas of decimal fractions (which become more and more important as the use of machines increases), percentage, and business math. The business knowledge should include familiarity with the cash register, adding machine, and calculator. With a thorough knowledge of these skills you should be mathematically ready to take a place in the hospitality field.

Solving a problem means finding a way out of a difficulty, a way around an obstacle, attaining an aim which was not immediately attainable. Solving problems is the specific achievement of intelligence, and intelligence is the specific gift of mankind: solving problems can be regarded as the most characteristic human activity...[1]

How long before success?

The answer is now. No one need fail this mathematics course.

How much math do you need to know for the foodservice/lodging industry? That, of course, is up to you. It depends on how far you want to go in the field. Do you plan to wash dishes, or be a busboy, all your life? Is being a waiter or waitress as far as you want to go in the food or service industry? Make no mistake about it—you can do any of those things mentioned above and earn enough to keep a family. However, the hospitality field offers many more opportunities for those who want to advance. For example, being a top chef in a major hotel can place you in a top income bracket. The entire field of management is also wide open to you. Or do you want to own your own place, be your own boss? Take your choice. After you have decided just where you want to go, you can answer the question about how much math you will need. You may need only the simplest math background, or you may need to become deeply involved in the higher mathematics. Certainly, whatever your choice, you will need the fundamentals—adding, subtracting, multiplying, and dividing. You will need to know how to use these processes or operations on whole numbers and fractions (including decimal fractions). Then, too, you will need to know the how and why of percentage. Depending on your career decisions you may need to know much more.

The emphasis here is on the word *know*. The real test of your knowledge will come when you can apply *what you know* on the job.

While this workbook will deal primarily with the kind of problems you will face on the job, don't forget that math is also fun. There are many good books devoted to mathematics for pleasure including some fairly inexpensive paperbacks. See for yourself how much fun math can be. You will be pleasantly surprised at the vast amount of knowledge you can gain while you are enjoying it.

---

[1] George Polya, *Mathematical Discovery 1* (New York: John Wiley and Sons, Inc., 1962).

FIT YOURSELF INTO THE FUTURE

CONTRACT

Contract

NAME_____

PERIOD_____ DATE_____

_____
SCHOOL

This agreement shall be binding upon the student and the teacher who are parties to the contract which is signed below.

This agreement shall award a GRADE to the student upon completion of the contract as stated below.

RESPONSIBILITY–TEACHER

Upon satisfactory completion of this contract the undersigned teacher agrees to award a grade as follows:

(1)_____

(2)_____

(3)_____

(4)_____

RESPONSIBILITY–STUDENT

In return for the grade listed above, the undersigned student agrees to complete the following assignment(s) on or before

(1)_____   _____

(2)_____   _____

(3)_____   _____

(4)_____   _____

(5)_____   _____

As proof that he/she has really completed the assignment(s), the student agrees to complete a written and/or oral examination, which shall test the power to use the skills involved in the assignments listed above, with a percentage score of no less than ____ percent. If the student completes the examination(s) with a percentage score lower than ____ percent, he/she agrees to repeat the assignment(s), and/or to negotiate a new contract.

NONCOMPLETION OF CONTRACT

If this contract is not completed as per the terms above, the teacher shall be released from all claims upon a grade by the student and the terms agreed upon become null and void. Notwithstanding, if, in the opinion of the teacher, circumstances warrant granting a passing grade, the teacher shall have the right and power to do so.

THIS AGREEMENT HAS BEEN REACHED THIS ____ DAY OF _____, 19 ___.

STUDENT   TEACHER
SIGNATURE_____ SIGNATURE_____

Permission is hereby granted to the teacher using this book to reproduce this page as needed.

ATTENDANCE RECORD

Keep a record of your class attendance on this chart. Make an "X" in the square that represents days that you do not attend this class. At the end of each month total the number of days you were absent and the number of days this class met. Then compute the percentage figure that represents the number of days you were absent.

Day	1	2	3	4	5	6	7	8	9	10	11	12	13	14	15	16	17	18	19	20	21	22	23	24	25	26	27	28	29	30	31	Number of class days	Days attended class	Days absent from class	Percentage of absence
September																																			
October																																			
November																																			
December																																			
January																																			
February																																			
March																																			
April																																			
May																																			
June																																			
July																																			
August																																			

Look at your record of attendance from the point of view of an employer. Could you afford to hire yourself? Why does the employer, be he the owner or the manager, have to be concerned with absence from the job?

Permission is hereby granted to the teacher using this book to reproduce this page as needed.

WEEKLY TEST RECORD

													QUARTER			SEMESTER	
Test Number	1	2	3	4	5	6	7	8	9	10	11	12	Totals	Averages	Letter Grade	Ave	Letter Grade
Number of Problems															✗		✗
Number Correct																	
Percent Correct																	

Kinds of problems missed—add, mult, sub, div, decimals, percent, etc.

EVALUATION OF MY WORK

Why I missed problems
1. _____
2. _____
3. _____
4. _____
5. _____
6. _____
7. _____
8. _____
9. _____
10. _____
11. _____
12. _____

What I will do to keep from making the same kind of errors again
1. _____
2. _____
3. _____
4. _____
5. _____
6. _____
7. _____
8. _____
9. _____
10. _____
11. _____
12. _____

Permission is hereby granted to the teacher using this book to reproduce this page as needed.

LIST OF SYMBOLS

ADDITION

+ means add, plus, sum, increased by, and.
(3 + 4) three plus four, the sum of three and four.

MULTIPLICATION

× or · or ( )( ) all mean times, multiply, product.
(3 × 4) three times four.
(3 · 4) the raised dot says three times four also.
(3)(4) the sets of parentheses say three times four too.
All of the above stand for the product of three multiplied by four.

SUBTRACTION

− means subtract, minus, take away, less, decreased by, difference.
(4 − 3) four take away three.

DIVISION

⌐ or − or ÷ all mean some form of division.
⌐ The division box is often called the "gazinta" (goes into) sign. 6⌐12 is read "12 divided by 6" but will usually be read by most people "6 goes into 12." Thus, the sign ⌐ is often jokingly called the "gazinta" symbol.
− (the fraction bar) is read "divided by." Every fraction is a division problem and stands for a quotient. Thus, $\frac{3}{4}$, or three-fourths, means "3 divided by 4" and is another way of expressing the quotient .75. "Quotient" is a name for the answer to a division problem.
÷ means "divided by" too. So, 12 ÷ 6 is read "twelve divided by six."

RATIO AND PROPORTION

Dictionaries define ratio as a way to express a quotient, so the expression of a ratio is the expression of a division problem. But so are fractions division problems. How are fractions and ratios different from each other? Ratios are usually thought of as comparisons or rates. For example, the cafe sells three times as many lunches as dinners. You can also say "dinners outsell lunches at the rate of 3 to 1." Rate may also apply to packaging, i.e., 6 cans to the case or 48 pounds to a unit.
: is read "is to," as 5:10 (5 is to 10).

− (the fraction bar) is read "is to" in ratio and proportion problems. Thus, $5:10 = \frac{5}{10}$ is read "five is to ten as five is to ten." (Notice that the equal sign here becomes "as.")

/ (the slash) is also used to express fractions, but appears to be gaining ground rapidly as the favored way of expressing a ratio. The slash is also read "is to." Thus, $5:10 = 1:2$ is read "five is to ten as one is to two" and $\frac{5}{10} = \frac{1}{2}$ and $5/10 = 1/2$ are all read exactly the same way. Proportion is the expression of an equality relationship between the terms of ratios. Thus, in the proportion $5/10 = 1/2$ the relationship of five to ten is exactly the same as the relationship of the one to the two, and the two ratios are equal.

EQUALITY AND INEQUALITY

= means equal, is equal to, is, as. $(2 + 2 = 3 + 1)$
≠ is not equal to. $(3 \neq 4)$
> is greater than. $(6 > 4)$
≯ is not greater than. $(3 + 1 \ngtr 5)$
< is less than. $(4 < 6)$
≮ is not less than. $(6 \nless 4)$
√ Radical sign, the square root of.
( ) Parentheses, do me first.
[ ] Brackets.
{ } Braces, the set of, the set whose members are.
% Percent.
L.C.D. Least Common Denominator.
L.C.M. Least Common Multiple.
G.C.D. Greatest Common Divisor.

EXPONENTS AND POWERS

base→3^2←Exponent or the second power.

The exponent is a small numeral written in the upper right corner of a base numeral and tells how many times the base numeral is used as a factor. A *factor* is one of the numbers used in multiplication to yield a product.

3^2 means 3×3 or 9 and is read "three squared" or "the square of three" or "three to the second power."

3^3 means $3 \times 3 \times 3$ or 27 and is read "three cubed" or "the cube of three" or "three to the third power."

3^4 means $3 \times 3 \times 3 \times 3$ or 81 and is read "three to the fourth power." This method is followed for exponents greater than three.

ADDITION TABLE

Use this table to compute addition problems or to discover number patterns.

PLUS +	0	1	2	3	4	5	6	7	8	9			
0	0	1	2	3	4	5	6	7	8	9			
1	1	2	3	4	5	6	7	8	9	10			
2	2	3	4	5	6	7	8	9	10	11			
3	3	4	5	6	7	8	9	10	11	12			
4	4	5	6	7	8	9	10	11	12	13			
5	5	6	7	8	9	10	11	12	13	14			
6	6	7	8	9	10	11	12	13	14	15			
7	7	8	9	10	11	12	13	14	15	16			
8	8	9	10	11	12	13	14	15	16	17			
9	9	10	11	12	13	14	15	16	17	18			

Note: This table has been left open on the right side and on the bottom to indicate that the table continues to infinity. You may use the blank squares to continue your own table. Remember, it can be extended just as far as you want to extend it.

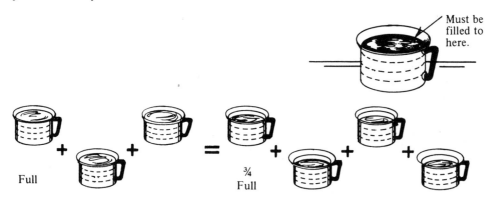

MULTIPLICATION TABLE I

Use this table to compute multiplication problems and to discover number patterns.

Times / FACTOR (top); FACTOR (left side)

X •	0	1	2	3	4	5	6	7	8	9		
0	0	0	0	0	0	0	0	0	0	0		
1	0	1	2	3	4	5	6	7	8	9		
2	0	2	4	6	8	10	12	14	16	18		
3	0	3	6	9	12	15	18	21	24	27		
4	0	4	8	12	16	20	24	28	32	36		
5	0	5	10	15	20	25	30	35	40	45		
6	0	6	12	18	24	30	36	42	48	54		
7	0	7	14	21	28	35	42	49	56	63		
8	0	8	16	24	32	40	48	56	64	72		
9	0	9	18	27	36	45	54	63	72	81		

You may continue to fill in this table. It is limited only by your imagination.

Note: The numbers indicated by the numerals on the marked diagonal are called squares. For example, 5^2 is read "five squared", or "five to the second power", and means that five is used as a factor two times, or 5×5, or 25.

All the numerals written inside the grid—those that are not factors—are answers to multiplication problems and are called *products*.

Squares are special products—they are the product of a factor multiplied by itself.

Note 2: Use this table for some division facts too. Find the dividend inside the body of the table, locate the divisor in the left-hand factor column, the quotient is in the top factor row above the dividend. For example, $72 \div 8$. Locate the row that contains the 8 and the 72. The quotient, 9, is directly above the 72 in the factor row.

MULTIPLICATION TABLE II

X	0	1	2	3	4	5	6	7	8	9	10	11	12	13	14	15	16	17	18	19	20	21	22	23
0	0	0	0	0	0	0	0	0	0	0	0	0	0	0	0	0	0	0	0	0	0	0	0	0
1	0	1	2	3	4	5	6	7	8	9	10	11	12	13	14	15	16	17	18	19	20	21	22	23
2	0	2	4	6	8	10	12	14	16	18	20	22	24	26	28	30	32	34	36	38	40	42	44	46
3	0	3	6	9	12	15	18	21	24	27	30	33	36	39	42	45	48	51	54	57	60	63	66	69
4	0	4	8	12	16	20	24	28	32	36	40	44	48	52	56	60	64	68	72	76	80	84	88	92
5	0	5	10	15	20	25	30	35	40	45	50	55	60	65	70	75	80	85	90	95	100	105	110	115
6	0	6	12	18	24	30	36	42	48	54	60	66	72	78	84	90	96	102	108	114	120	126	132	138
7	0	7	14	21	28	35	42	49	56	63	70	77	84	91	98	105	112	119	126	133	140	147	154	161
8	0	8	16	24	32	40	48	56	64	72	80	88	96	104	112	120	128	136	144	152	160	168	176	184
9	0	9	18	27	36	45	54	63	72	81	90	99	108	117	126	135	144	153	162	171	180	189	198	207
10	0	10	20	30	40	50	60	70	80	90	100	110	120	130	140	150	160	170	180	190	200	210	220	230
11	0	11	22	33	44	55	66	77	88	99	110	121	132	143	154	165	176	187	198	209	220	231	242	253
12	0	12	24	36	48	60	72	84	96	108	120	132	144	156	168	180	192	204	216	228	240	252	264	276
13	0	13	26	39	52	65	78	91	104	117	130	143	156	169	182	195	208	221	234	247	260	273	286	299
14	0	14	28	42	56	70	84	98	112	126	140	154	168	182	196	210	224	238	252	266	280	294	308	322
15	0	15	30	45	60	75	90	105	120	135	150	165	180	195	210	225	240	255	270	285	300	315	330	345
16	0	16	32	48	64	80	96	112	128	144	160	176	192	208	224	240	256	272	288	304	320	336	352	368
17	0	17	34	51	68	85	102	119	136	153	170	187	204	221	238	255	272	289	306	323	340	357	374	391
18	0	18	36	54	72	90	108	126	144	162	180	198	216	234	252	270	288	306	324	342	360	378	396	414

SUBTRACTION TABLE

Minus					MINUEND								
−	0	1	2	3	4	5	6	7	8	9			
0	0	1	2	3	4	5	6	7	8	9			
1		0	1	2	3	4	5	6	7	8			
2			0	1	2	3	4	5	6	7			
3				0	1	2	3	4	5	6			
4					0	1	2	3	4	5			
5						0	1	2	3	4			
6							0	1	2	3			
7								0	1	2			
8									0	1			
9										0			

(Left-side label: SUBTRAHEND)

Note: Be sure to read the name of the top row of numerals and the column on the left-hand side. Associate those names with subtraction problems.

Why are some of the squares on the grid left blank? What kind of numbers do you have to know about before you can fill in the blank spaces?

The blank spaces in the grid above really demonstrate why we say that subtraction is not "closed."

Can you extend the table above? Do you see and recognize the number patterns above?

EQUIVALENT FRACTIONS

Read the fractions across the rows. All fractions in a row are equivalent to all other fractions in that row. That means that they name the same number or that they are located at the same spot on the number line.

The first fraction listed on the left, in each case is in lowest, or simplest, terms. All fractions that follow the first one are *multiples* of the first fraction. All fractions that follow the first fraction can be *reduced* to the first fraction.

Study the lists provided. Look for patterns. Note that the equivalent fractions listed are not complete. The lists could be continued endlessly.

$\frac{1}{2}, \frac{2}{4}, \frac{3}{6}, \frac{4}{8}, \frac{5}{10}, \frac{6}{12}, \frac{7}{14}, \frac{8}{16}, \frac{9}{18}, \frac{10}{20}, \frac{11}{22}, \frac{12}{24}, \frac{13}{26}, \frac{14}{28}, \frac{15}{30}, \frac{16}{32}, \frac{17}{34}, \frac{18}{36}, \frac{19}{38}, \frac{20}{40} \ldots$

$\frac{1}{3}, \frac{2}{6}, \frac{3}{9}, \frac{4}{12}, \frac{5}{15}, \frac{6}{18}, \frac{7}{21}, \frac{8}{24}, \frac{9}{27}, \frac{10}{30}, \frac{11}{33}, \frac{12}{36}, \frac{13}{39}, \frac{14}{42}, \frac{15}{45}, \frac{16}{48}, \frac{17}{51}, \frac{18}{54}, \frac{19}{57}, \frac{20}{60} \ldots$

$\frac{1}{4}, \frac{2}{8}, \frac{3}{12}, \frac{4}{16}, \frac{5}{20}, \frac{6}{24}, \frac{7}{28}, \frac{8}{32}, \frac{9}{36}, \frac{10}{40}, \frac{11}{44}, \frac{12}{48}, \frac{13}{52}, \frac{14}{56}, \frac{15}{60}, \frac{16}{64}, \frac{17}{68}, \frac{18}{72}, \frac{19}{76}, \frac{20}{80} \ldots$

$\frac{1}{5}, \frac{2}{10}, \frac{3}{15}, \frac{4}{20}, \frac{5}{25}, \frac{6}{30}, \frac{7}{35}, \frac{8}{40}, \frac{9}{45}, \frac{10}{50}, \frac{11}{55}, \frac{12}{60}, \frac{13}{65}, \frac{14}{70}, \frac{15}{75}, \frac{16}{80}, \frac{17}{85}, \frac{18}{90}, \frac{19}{95}, \frac{20}{100} \ldots$

$\frac{1}{6}, \frac{2}{12}, \frac{3}{18}, \frac{4}{24}, \frac{5}{30}, \frac{6}{36}, \frac{7}{42}, \frac{8}{48}, \frac{9}{54}, \frac{10}{60}, \frac{11}{66}, \frac{12}{72}, \frac{13}{78}, \frac{14}{84}, \frac{15}{90}, \frac{16}{96}, \frac{17}{102}, \frac{18}{108}, \frac{19}{114}, \frac{20}{120} \ldots$

$\frac{1}{7}, \frac{2}{14}, \frac{3}{21}, \frac{4}{28}, \frac{5}{35}, \frac{6}{42}, \frac{7}{49}, \frac{8}{56}, \frac{9}{63}, \frac{10}{70}, \frac{11}{77}, \frac{12}{84}, \frac{13}{91}, \frac{14}{98}, \frac{15}{105}, \frac{16}{112}, \frac{17}{119}, \frac{18}{126} \ldots$

$\frac{1}{8}, \frac{2}{16}, \frac{3}{24}, \frac{4}{32}, \frac{5}{40}, \frac{6}{48}, \frac{7}{56}, \frac{8}{64}, \frac{9}{72}, \frac{10}{80}, \frac{11}{88}, \frac{12}{96}, \frac{13}{104}, \frac{14}{112}, \frac{15}{120}, \frac{16}{128}, \frac{17}{136}, \frac{18}{144} \ldots$

$\frac{1}{9}, \frac{2}{18}, \frac{3}{27}, \frac{4}{36}, \frac{5}{45}, \frac{6}{54}, \frac{7}{63}, \frac{8}{72}, \frac{9}{81}, \frac{10}{90}, \frac{11}{99}, \frac{12}{108}, \frac{13}{117}, \frac{14}{126}, \frac{15}{135}, \frac{16}{144}, \frac{17}{153}, \frac{18}{162} \ldots$

$\frac{1}{10}, \frac{2}{20}, \frac{3}{30}, \frac{4}{40}, \frac{5}{50}, \frac{6}{60}, \frac{7}{70}, \frac{8}{80}, \frac{9}{90}, \frac{10}{100}, \frac{11}{110}, \frac{12}{120}, \frac{13}{130}, \frac{14}{140}, \frac{15}{150}, \frac{16}{160}, \frac{17}{170} \ldots$

$\frac{1}{11}, \frac{2}{22}, \frac{3}{33}, \frac{4}{44}, \frac{5}{55}, \frac{6}{66}, \frac{7}{77}, \frac{8}{88}, \frac{9}{99}, \frac{10}{110}, \frac{11}{121}, \frac{12}{132}, \frac{13}{143}, \frac{14}{154}, \frac{15}{165}, \frac{16}{176}, \frac{17}{187} \ldots$

$\frac{1}{12}, \frac{2}{24}, \frac{3}{36}, \frac{4}{48}, \frac{5}{60}, \frac{6}{72}, \frac{7}{84}, \frac{8}{96}, \frac{9}{108}, \frac{10}{120}, \frac{11}{132}, \frac{12}{144}, \frac{13}{156}, \frac{14}{168}, \frac{15}{180}, \frac{16}{192}, \frac{17}{204} \ldots$

$\frac{1}{13}, \frac{2}{26}, \frac{3}{39}, \frac{4}{52}, \frac{5}{65}, \frac{6}{78}, \frac{7}{91}, \frac{8}{104}, \frac{9}{117}, \frac{10}{130}, \frac{11}{143}, \frac{12}{156}, \frac{13}{169}, \frac{14}{182}, \frac{15}{195}, \frac{16}{208}, \frac{17}{221} \ldots$

PRIME NUMBERS

Factors are the numbers you multiply together to get a product. A problem can have two or more factors. (For example: 2 × 3 = 6. The 2 and the 3 are factors of the problem. 2 × 3 × 6 = 36. In this problem, the 2, 3, and 6 are all factors of the problem.)

Prime Numbers are numbers that have two distinct or different factors; prime numbers can only be divided evenly by themselves and 1. However, the number 1 is not considered a prime number. (For example: 7 is a prime number. The factors of 7 are 1 and 7. The only numbers that will divide evenly into 7 are 1 and 7.)

However, the number 36 used in the example above has many factors. The factors of 36 are 1, 2, 3, 4, 6, 9, 12, 18, and 36. All of these factors will divide evenly into 36. Can you discover all the multiplication combinations that can be written using the factors above to get a product of 36?

TABLE OF PRIME NUMBERS FROM 1 TO 1000

2	3	5	7	11	13	17	19	23	29
31	37	41	43	47	53	59	61	67	71
73	79	83	89	97	101	103	107	109	113
127	131	137	139	149	151	157	163	167	173
179	181	191	193	197	199	211	223	227	229
233	239	241	251	257	263	269	271	277	281
283	293	307	311	313	317	331	337	347	349
353	359	367	373	379	383	389	397	401	409
419	421	431	433	439	443	449	457	461	463
467	479	487	491	499	503	509	521	523	541
547	557	563	569	571	577	587	593	599	601
607	613	617	619	631	641	643	647	653	659
661	673	677	683	691	701	709	719	727	733
739	743	751	757	761	769	773	787	797	809
811	821	823	827	829	839	853	857	859	863
877	881	883	887	907	911	919	929	937	941
947	953	967	971	977	983	991	997		

ESSENTIALS OF MATHEMATICS

DIAGNOSTIC TESTS

WHOLE NUMBERS

Directions: Solve the problems. Make an answer column on the right-hand side of your paper. Show all your work.

Add:

A.	2457	B.	909	C.	234	D.	123	E. $.06 + $1.97 + $22.07 =	F.	31
	+3321		+704		987		439			206
					651		216			7895
										432
										89
										4

Subtract:

G.	7698	H.	7698	I.	7007	J.	7606	K.	$10.37	L.	$6.00
	−4377		−5479		−1269		−1089		−9.99		−.07

M. $300 − $15.15 =

Multiply:

N.	342	O.	3426	P.	9876	Q.	909	R.	7006	S.	567	T.	4067	U.	4607
	×2		× 3		×7		×8		× 8		×10		×11		×204

V. 567
   ×894

Divide:

W. 2)26840   X. 49 ÷ 23 =   Y. 9)10893

FRACTIONS

Directions: Solve the problems. Make an answer column on the right-hand side of your paper. Show all your work.

Raise to higher terms:

A. $\dfrac{1}{2} = \dfrac{}{4} = \dfrac{}{8} = \dfrac{}{16}$

Express in lowest terms:

B. $\dfrac{6}{8} = \dfrac{}{}$ C. $\dfrac{18}{30} = \dfrac{}{}$

19

Add—Express answers in lowest terms:

D. $\dfrac{1}{3}+\dfrac{1}{3}=$ E. $\dfrac{1}{2}+\dfrac{1}{2}=$ F. $\dfrac{1}{6}+\dfrac{2}{6}=$ G. $\dfrac{1}{2}+\dfrac{1}{3}=$ H. $\dfrac{3}{4}+\dfrac{2}{4}=$ I. $\dfrac{1}{2}$ J. $\dfrac{1}{3}$
$\dfrac{1}{3}$ $\dfrac{1}{6}$
$\dfrac{3}{4}$ $\dfrac{1}{9}$

Express as a mixed number:

K. $\dfrac{10}{3}=$ L. $\dfrac{30}{24}=$ M. $\dfrac{17}{9}=$

Check—Are the fractions in lowest terms?

Subtract:

N. $\dfrac{2}{3}-\dfrac{1}{3}=$ O. $\dfrac{10}{12}-\dfrac{4}{12}=$ P. $\dfrac{2}{3}-\dfrac{1}{6}=$ Q. $\dfrac{9}{16}$
$-\dfrac{1}{3}$

Mixed numbers—Add or subtract:

R. $1\dfrac{1}{4}+\dfrac{2}{4}=$ S. $1\dfrac{1}{4}$ T. $3\dfrac{1}{2}$ U. $3\dfrac{5}{8}$ V. $5\dfrac{7}{12}$
$-\dfrac{2}{4}$ $+2\dfrac{5}{8}$ $+\dfrac{3}{4}$ $+2\dfrac{2}{3}$

Multiply:

W. $\dfrac{1}{3}\times\dfrac{1}{2}=$ X. $\dfrac{2}{3}\times\dfrac{3}{4}=$

Divide:

Y. $\dfrac{1}{3}\div\dfrac{1}{2}=$

DECIMALS

Directions: Solve the problems. Make an answer column on the right-hand side of your paper. Show all your work. CAUTION—Be sure your decimal points can be easily seen so that there can be no doubt that you know how to place them.

Write the decimal equivalent of the following common fractions:

A. $\dfrac{1}{4}=$ B. $\dfrac{1}{2}=$ C. $\dfrac{1}{3}=$ D. $\dfrac{1}{5}=$ E. $\dfrac{1}{8}=$ F. $\dfrac{3}{4}=$ G. $\dfrac{2}{3}=$ H. $\dfrac{1}{16}=$

Add:

I. $.2+.3=$ J. $.7+.9=$ K. $.2$ L. 2.3 M. 2.4 N. $3.2+33.5+4.9+.7=$
$.3$ $+1.5$ 3.5
$.1$ 6.7
$.2$
$.1$

O. $1.34 + $2.49 + $34.98 + $16.77 =

P. .24
    1.36
    3.98

Q. .2596
    .3600
    .4970
    .1234

R. .4321 + .39 + .298 + .001 =

Subtract:

S. .9 − .4 =

T. 1.3
    − .9

U. 1.9 − .9 =

V. 1.7 − .07 =

W. Take .007 from 1.2.

X. $1.34
    − .98

Y. $23.37
    − 11.79

Multiply:

Z. .1 × .1 =

AA. .3 × .4 =

BB. 3.98
    × 4

CC. 34.39
    × .21

DD. .78903
    × .976

EE. .75
    × .75

FF. .705
    × .03

Divide:

GG. .12 ÷ 3 =

HH. 1.2 ÷ 3 =

II. 12 ÷ .3 =

JJ. 3.3$\overline{)6.6}$

KK. .33$\overline{)6.93}$

LL. .07$\overline{).4949}$

MM. 9$\overline{).8199}$

NN. 327$\overline{)981.1308}$

OO. 7$\overline{).007}$

PP. .003$\overline{)3}$

Round off to the nearest hundredth:

QQ. .199 =

RR. .2023 =

SS. .1373 =

TT. .0098 =

UU. $.095 =

PERCENTAGE

Directions: Solve the problems. Make an answer column on the right-hand side of your paper. Show all your work.

Select the correct name:

A. 25%

B. 50%

C. 33$\frac{1}{3}$%

D. $\frac{1}{4}$%

E. 6%

1. .5

2. $\frac{3}{50}$

3. $\frac{1}{3}$

4. $\frac{1}{4}$

5. .0025

Continued on next page.

Find the missing number or percent:

F. 6% of 12 =

G. 5% of 100 =

H. 10% of $200 =

I. 3% of what number = 27

J. 15% of x = 45

K. $33\frac{1}{3}$% of ☐ = 3

L. What percent of 36 is 3.6?

M. ?% of 16 = 4

N. ☐% of 600 = 36

O. 16 is what percent of 32?

P. 4 = ☐% of 12?

Q. 8 = 50% of ☐?

R. $\frac{5}{100} = \frac{x}{800}$

S. $\frac{7}{100} = \frac{70}{x}$

T. $\frac{☐}{100} = \frac{4}{16}$

U. Beverly saves 10% of her earnings. Last week she earned $50 plus $70 in tips. How much will she put in the bank?

V. Meryl gives her parents 35% of her salary to pay some of her room and board. She earns $200 a month. How much does she give her parents?

W. Don and Mike agree to have their pay adjusted every 6 months in accordance with the semiannual profit report. If profits go up, their pay increases. If profits go down, so does their pay. They start working January 1 earning $400 a month. On June 1 they receive a 5% increase in salary. The next January they get a 5% decrease. How much more or less are Don and Mike earning at the start of their second year? (Be sure to label your answer either "more" or "less.")

NUMBER AND NUMERAL

The *digits* are 0, 1, 2, 3, 4, 5, 6, 7, 8, and 9. These represent all the numerals we write in any one column in our number system. All other numerals are made by combining digits, such as 10, 21, and 123.

We write numerals to represent the numbers we think about. We associate the numeral with a number in our minds. So we think of "nothing" and we write a zero (0). It can be said that the zero indicates an absence of any other digit. It has also been said the zero holds a place for other digits. This is true, but it tells us also that nothing is there. If, for example, you are taking inventory, and come to an empty space on the shelf that usually holds canned tomatoes, you mark down a zero on the inventory sheet. That zero does two things (or may do two things): it shows that there are no more canned tomatoes, and it may hold the place for tomatoes until another shipment comes in. After the shipment is received, and the cans have been placed on the shelf, and the space is no longer empty, the zero can be replaced with another digit or digits.

When we write the numeral "1," we associate it with a single item or thing, like one steak, one dinner, one dollar, one car, or one person.

When we write the numeral "2," we associate it with two things, like two steaks, dinner for two people, and so on. The idea of associating numerals with the numbers we think about makes life much easier for us. What do you think the term "twoness" means?

Now you go over all the remaining digits and answer your own questions about "threeness," "fourness," and so forth.

PLACE VALUE

Place value is the innovation that makes our number system far superior to other number systems. Place value makes it possible for our system to use only ten digits to express quantities representing numbers from the very lowest to the highest.

The digits, as stated above, are 0, 1, 2, 3, 4, 5, 6, 7, 8, and 9. Each digit has a value of its own. Yet each digit can assume many different values depending on the column it occupies. Thus, the digit 3 can stand for three things or 30 things or three-tenths of a thing or 3 million things. Each digit can represent many numerical values because of place value.

Our system is a decimal, or base ten, system. The word "decimal" comes from the Latin word *decem* and means "ten." Our system is based on the number 10 and the powers of 10.

Look at the chart below to understand the directions that follow.

	10^9	10^8	10^7	10^6	10^5	10^4	10^3	10^2	10^1	10^0	• (decimal point) and	10^{-1}	10^{-2}	10^{-3}	10^{-4}	10^{-5}	10^{-6}	10^{-7}	10^{-8}	10^{-9}
							1000	100	10	1		$\frac{1}{10}$	$\frac{1}{100}$	$\frac{1}{1000}$						
	billions	hundred millions	ten millions	millions	hundred thousands	ten thousands	thousands	hundreds	tens	ones (units)		tenths	hundredths	thousandths	ten thousandths	hundred thousandths	millionths	ten millionths	hundred millionths	billionths

line

	10^9	10^8	10^7	10^6	10^5	10^4	10^3	10^2	10^1	10^0	.	10^{-1}	10^{-2}	10^{-3}	10^{-4}	10^{-5}	10^{-6}	10^{-7}	10^{-8}	10^{-9}
a										9	.									
b									6		.									
c								5			.									
d							3				.									
e						1					.									
f					0						.									
g				2							.									
h			4								.									
i		7									.									
j	8										.									
k								1	0	1	.									
l						2	1	4	6	7	.									
m				1	0	0	1	0	3	2	.									
n											.	1								
o											.	0	2							
p											.	0	0	5						
q											.	1	0	7						
r											.	2	4	7	9	6				
s									1	1	.	2	1							
t							3	4	5	8	.	9	8	6						
u						4	0	0	6	7	.	0	0	7	6					
v	1	0	0	0	0	0	0	0	0	0	.	0	0	0	0	0	0	0	0	1
w																				
x																				
y																				
z																				

Read line a. It says "nine" and since the 9 is in the one's column it represents 9 × 1, or 9. Now read line b. It says "six" but since the 6 is in the ten's column it represents 6 × 10, or 60. If numerals were written in labelled columns we could communicate the number described. However, since numbers are written with no reference to column values, the writer and the reader both must be able to understand what is said. So, instead of writing "6 tens," we write "60," and both the writer and those reading the numeral understand that the number idea is "six tens and no ones" or "sixty." Communication is simplified.

Line c then communicates the idea of 5×10^2, or $5 \times 10 \times 10$, or 5×100, or five hundred. When it is written this way, 500, it says 5 hundreds plus no tens plus no ones.

Line k reads "one hundred one" and is the combination of 1 hundred plus 0 tens plus 1 one.

The numbers on the left-hand side of the decimal point represent the <u>whole numbers.</u> The numerals are in groups of three digits separated by a comma, and each group of three is called a period in mathematics. Each period has the name of a major category, such as billion, million, and so forth. Thus, 1,001,001,001 is read 1 billion, 1 million, 1 thousand, 1. So, if you can read numbers to nine-hundred ninety-nine (999), you can read any number in our decimal system.

Read the number in each period. Read the name of the column just to the right of the comma each time you come to a comma. The number shown below reads "nine billion, ninety-nine million, three hundred twenty-four thousand, six hundred fifty-seven."

```
   ┌─BILLION
   │   ┌────MILLION
   │   │     ┌────────THOUSAND
   ↓   ↓     ↓
   9,099,324,657
```

The <u>decimal point</u> separates whole numbers from fractional numbers when the fraction is expressed in decimal form. The decimal point is the only place where the word "and" is correctly read in our number system. The word "and" means a fraction comes next. The place value of decimal fractions is shown on the place value chart. The values move to the right from the decimal point in decreasing value.

$$\left(\frac{1}{10}, \frac{1}{100}, \frac{1}{1000}, \frac{1}{10,000}, \frac{1}{100,000}, \frac{1}{1,000,000}\right)$$

Thus, line n on the chart reads "one-tenth," line o reads "two hundredths," and line q reads "one hundred seven thousandths." The fraction takes the name of the last column on the right in which a digit appears. So, line r reads "twenty-four thousand, seven hundred ninety-six hundred thousandths."

Now, note the use of the word "and" to signify the decimal point. Read line s as "eleven and twenty-one hundredths." Read line t "three thousand, four hundred fifty-eight and nine hundred eighty-six thousandths." Try to read the other lines on the chart. Fill in and read some numbers of your own.

NUMBER LINES

A number line goes on forever. We say it is infinitely long. The number line represented here is called a positive number line.

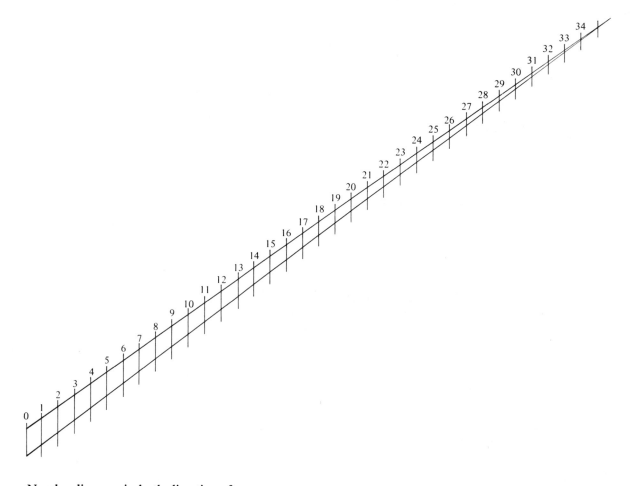

Number lines go in both directions from zero.

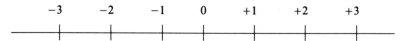

When the number line moves to the right of zero, the numbers corresponding to the marks on the number line are said to be positive (+). When the number line moves to the left of zero, it is said to be negative (−). Zero is the halfway point between −1 and +1 and is called the midpoint of the number line.

HOW TO ADD

Addition in mathematics is the process of uniting two numbers to obtain a third number. In modern mathematics you would substitute the word "quantity" for "number" and have a definition reading something like this: addition is the operation of combining two or more quantities to obtain a third quantity called the <u>sum</u>. The numbers, or quantities, to be added are called <u>addends.</u> For example:

ADDEND	PLUS	ADDEND	EQUALS	SUM
3	+	4	=	7

```
 3  ADDEND
 4  ADDEND
+5  ADDEND
12  SUM
```

The operation of addition makes use of the plus, or addition, sign (+).

If you are to go to work in the foodservice or hospitality field, you will have great need to know all the addition combinations and should make a strong effort to memorize them. (These combinations and other tables are provided in this book.) In the future the foodservice/lodging industry may provide for computerized addition, and all the other operations as well, for most of their employees. Most cash registers can be used as adding machines, and many of the places where you will work will provide adding machines and/or calculators.

In order to do your best to serve your employer you should probably learn to use business machines, but don't let the fact that machines can do the mathematics operations fast and accurately make you think it isn't necessary for you to know and understand mathematical operations. You will undoubtedly find mathematical knowledge to be quite valuable to you. In addition, you will find that trouble comes from not knowing the combinations, from incorrect reading of the numerals, and from improper carrying.

When you add a single digit column or row there are few problems. Following is a horizontal addition problem:

$2+3+4+5+6+7=27$

The same problem stated in vertical form looks like this:

```
 2
 3
 4
 5
 6
 7
27
```

Note that the plus sign (+) is not necessary when more than two numerals are shown in a column.

$$\begin{array}{r} 3 \\ +2 \\ \hline 5 \end{array}$$

However, when you use only two numbers in a column, you need an operation sign.

When you are sure you know the addition combinations and have practiced with single digit numbers in rows and columns to prove your command of this knowledge, you can tackle numbers with digits in the ones, tens, hundreds, and other columns. This can still be simple addition which involves no carrying, or more complicated addition which involves carrying.

There is no carrying involved. Begin all adding in the one's column, move to the ten's column, then the hundred's column, and so on.

Step one.   Add the 4 and 2 and write the answer 6 in the sum.

Step two.   Move to the ten's column and add the 3 and 3. Write the answer 6, meaning 6 tens, in the ten's place in the sum.

Step three.   Move to the hundred's column and add the 2 to the 4 and write the answer, 6, in the hundred's place in the sum. Note that this digit stands for 6 hundreds or 600.

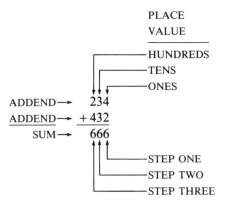

When problems require carrying, at least one more move is involved.

Step one.   Add the 4 and 7 and get the answer 11 or 1 ten and 1 one. Place the 1 one in the one's column in the one's place in the sum. <u>Carry</u> the other 1 (1 ten) to the top of the ten's column.

Step two.   Next add the ten's column, including the carried digit. (The ten's column now has three digits, a one, a three, and an eight.) Adding gives a partial sum of twelve (12) and the digits stand for 1 hundred and 2 tens. We have added 1 ten, 3 tens and 8 tens to get an answer of 12 tens (120). Place the 2 in the ten's column and carry the 1 to the top of the hundred's column.

Step three.   Finally (in this problem) add the three digits in the hundred's column, 1 + 2 + 9, and write the 12 in the hundred's and thousand's columns. The 2 goes in the hundred's column and, since there is no thousand's column in this problem, the 1 goes into the thousand's place in the sum.

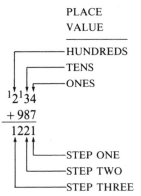

Now look at the same problem this way:

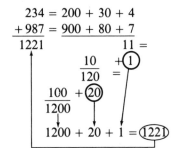

ADD ONES (UNITS)
10 + 1(REGROUPING).
MOVE 10 TO TEN'S COLUMN.
ADD TENS AND REGROUP.
ADD HUNDREDS
AND COMBINE
WITH TEN'S AND
ONE'S COLUMN.

Another troublemaker in addition is the zero. The zero is a place holder. It means the place is empty; that nothing is in that place.

In the number 302, each digit fills a place. The numeral 3 is in the hundred's place and means 3 hundreds; the numeral 0 (zero) is in the ten's place and means no tens; the numeral 2 is in the one's place and means 2 ones. The combined numerals stand for our idea of three hundred two.

Where you see the zero, you know it holds the place; that it means nothing is there in that particular place at this particular time or in this particular number. Now, see how to treat the zero when it shows up in addition.

```
 20
203
947
```

Start in the one's column. Add 3 ones to 7 ones, get 10 ones which is 1 ten and no ones. Write a zero in the one's column (standing for no ones) and carry (or tote) the 1 to the ten's column.

```
  1
 20
203
947
  0
```

Now add the digits in the ten's column and get seven. (1 + 2 = 3; 3 + 0 = 3; 3 + 4 = 7). Write the numeral 7 in the ten's column. Is there any numeral to carry?

```
  1
 20
203
947
 70
```

Finally, add the numerals in the hundred's column. (2 + 9 = 11). Write your answer in the hundred's column and the thousand's column. Actually the 1 in the thousand's column was carried and added to the zero in the thousand's column before writing the answer.

```
   1
  20
 2 03
 9 47
11 70
```

Study this method of adding. Do you get the same answer? Do you see what happened to the tote, or carry, numerals?

```
        20
       203
       947
      0 10
      0 7
       11
SUM = 1170
```

Here is another way to add:

```
  20 =          20 +  0
 203 =     200 + 0 +  3
 947 = (+)900 + 40 + 7
1170 =    1100 + 60 + 10
```

COMBINE

1160 + 10

COMBINE

1170

Modern mathematics books do this kind of addition under the title "Expanded Notation." Expanded notation is especially helpful in understanding other base operations as well as understanding our base ten system. In our problem above, we need to understand each step of the problem. Start by writing the names of the columns as below. (Remember 10^2 means 10 times 10 and is another way to write 100.) Write the problem.

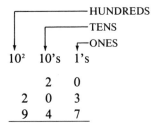

```
10²  10's  1's
      2    0
 2    0    3
 9    4    7
```

In the first line of the problem the 2 stands for 2 × 10 (it is in the ten's column and means there are 2 tens) and the zero (0) stands for no ones. So in expanded notation 20 is the sum of (2 × 10) plus (0 × 1) or

20 + 0 = 20.

Similarly, the other lines represent

(2 × 100) + (0 × 10) + (3 × 1)
   200   +   0   +   3
(9 × 100) + (4 × 10) + (7 × 1)
   900   +   40   +   7

Now take the numbers this way:

           (2 × 10) + (0 × 1)
(2 × 100) + (0 × 10) + (3 × 1)
(9 × 100) + (4 × 10) + (7 × 1)

Combine all the groups representing one column and you have

(2 × 100) + (9 × 100) = (11 × 100)
(2 × 10) + (0 × 10) + (4 × 10) = (6 × 10)
(0 × 1) + (3 × 1) + (7 × 1) = (10 × 1)

Take all the sums and combine them

(11 × 100) + (6 × 10) + (10 × 1) =
   1100   +   60   +   10   = 1170

You notice that you arrive at the same answer no matter which system you use. Expanded notation is very useful in understanding what happens.

There are many other methods that are useful. It is not necessary that they be detailed here. However, it is necessary that you, the learner, understand that it is possible for you to expand your mathematical knowledge by referring to many math books. Take a second look at that book you feel you hate; it may have something to tell you.

HOW TO MULTIPLY

1. Learn your tables, at least through the ten's. (See pages 12 and 13.)
2. Know the rules which make multiplying by 10, 100, 1000, and so on, easy. (See pages 35 and 36.)
3. When multiplying by more than one digit, solve for a partial product (partial answer) for each digit in the multiplier and add the partial products to get the product (answer).

Examples

Multiplying by one digit:

$$\begin{array}{r} 123 \\ \times 2 \\ \hline 246 \end{array}$$ (MULTIPLICAND OR FACTOR) (MULTIPLIER OR FACTOR) (PRODUCT)

Step one. (one's column)
$3 \times 2 = 6$

Step two. (ten's column)
$2 \times 2 = 4$

Step three. (hundred's column)
$1 \times 2 = 2$

When a zero is involved:

$$\begin{array}{r} 1203 \\ \times 3 \\ \hline 3609 \end{array}$$

Step one. (one's column)
$3 \times 3 = 9$

Step two. (ten's column)
$0 \times 3 = 0$

Step three. (hundred's column)
$2 \times 3 = 6$

Step four. (thousand's column)
$1 \times 3 = 3$

When carrying is involved:

$$\begin{array}{r} 1\,1 \\ 123 \\ \times 6 \\ \hline 738 \end{array}$$

Step one. (one's column) Multiply 3 × 6 = 18. Write the 8 under the 3 and carry the 1 to the ten's column.

Step two. (ten's column) Multiply 2 × 6 = 12; add the one carried to get 13. Write the 3 under the 2 and carry the 1 to the hundred's column.

Step three. (hundred's column) Multiply 1 × 6 = 6; add the one carried and get 7. Write the 7 under the 1. Read the product 738.

When carrying and a zero are involved:

```
 1 2
1203
 ×7   (3 × 7 = 21)
8421
```

Step one. (one's column) Multiply 3 × 7 = 21. Write the 1 under the 3 and carry the 2 to the ten's column.

Step two. (ten's column) Multiply 0 × 7 = 0; add the 2 to get a sum of 2. Write the 2 below the zero.

Step three. (hundred's column) Multiply the 2 × 7 = 14. There is no carried numeral to add, so write the 4 under the 2 and carry the 1 to the thousand's column.

Step four. (thousand's column) Multiply 1 × 7 = 7, add the one carried = 8, and write this sum under the 1. Notice how the numerals in the product are related to the column you are working with in the multiplicand.

Multiplying by two or more digits:

```
   ⑤⑥④
   ⑥⑥④
   4675  MULTIPLICAND
    ×89  MULTIPLIER
  42075  PARTIAL PRODUCT
 374000  PARTIAL PRODUCT
 416075  PRODUCT
```

Step one. Multiply each digit in the multiplicand by 9. Start on the right. Move left. Carry and add as before.

5 × 9 = ④5 Carry circled digit.
7 × 9 = 63 + 4 = ⑥7
6 × 9 = 54 + 6 = ⑥0
4 × 9 = 36 + 6 = 42
42075 is the partial product of 4675 × 9.

Step two. Multiply each digit in the multiplicand by 8 (the 8 represents 8 tens or 80). First, write a zero in the one's place in the partial product. (This is the answer to 0 × 8. Since you are now multiplying by 8 tens or 80 or (8 × 10), and if you multiply 4675 × 10 you get 46750, you would then be ready to multiply by 8 and your right-hand digit in the partial product would be the answer to 0 × 8 or 0.) Next, multiply all the digits in the multiplicand by 8 and get the second partial product. Carry the digits in squares. Remember, start at the right, move left.

5 × 8 = ④0
7 × 8 = 56 + 4 = ⑥0
6 × 8 = 48 + 6 = ⑤4
4 × 8 = 32 + 5 = 37

Step three. Add the partial products. The product is 416,075.

When zeros are involved with two-digit multiplication:

```
    ④⑥④
    ⑤⑥④
   46075
  ×   89
   41 4675
  368 6000
  410 0675
```

Step one. If zero or zeros are in the multiplicand, treat them exactly as you did in one-digit multiplication.

```
        ⑤⑥④
        ⑥⑥④
        4675
        8009
       42075  PAR. PROD.
   37400000  PAR. PROD.
   37442075  PRODUCT
```

Step two. If zero or zeros are in the multiplier, increase the partial product by short multiplication by adding zeros. First, multiply each digit in the multiplicand by 9. The amount left to multiply by is 8000 (8009-9). Place three zeros on the right-hand side of the second partial product. Multiply each digit (right to left) of the multiplicand by 8 and write the second partial product.

Step three. Add the partial products and write the product.

Multiplication is an extension of addition. It is an easy way to add groups that represent the same quantity. However, it is true that all multiplication problems can be solved by adding.

Do you know your multiplication tables? As in addition, it will make your job easier to learn those tables you do not know. Refer to the table and familiarize yourself with it so you can make quick reference to it when you need to use it. Do it now.

While multiplication may be looked upon as an easy way to add, there are some differences. Chiefly, the multiplication of a number is involved with adding an amount (or a quantity) to itself. If the numerals represent different quantities, they should probably be added. If they represent the same quantity, they should be multiplied. For example:

```
Add:          Add:          Multiply:      Multiply:
(Ex. 1)  17   (Ex. 2)  13   (Ex. 3)  13    (Ex. 4)  13 · 4 = 52
          7             13            ×4
         19             13            52
          3             13
          6             13
         52             52
```

While you can find a way to multiply the first example, it is probably easier to add. How would you regroup so you can multiply? (17 + 3 + 6 = 26 and 7 + 19 = 26. 26 · 2 = 52.) However, it is probably easier to multiply in example three rather than add as in example two. Example four is a multiplication problem stated using a different symbol. The point that it is easier to add in these examples than to multiply may be debatable. However, it is not debatable that some problems can be found which are harder to add than to multiply.

For example, try adding 5768 two hundred thirty-four times. Yet it is relatively easy to solve this problem by multiplication. Think of the job just writing 5768 two hundred thirty-four times! Even with an adding machine, the task takes time. Try it.

To multiply, the problem looks like this:

```
  5768
×  234
```

or

5768 · 234 =

Solving the problem requires that you take 5768 four times, then 30 times, then 200 times getting three partial products which you combine (add) for the final answer (product).

```
MULTIPLICAND       5768
MULTIPLIER          234
PARTIAL PRODUCT   23072    = 5768 four times
PARTIAL PRODUCT  173040    = 5768 thirty times
PARTIAL PRODUCT 1153600    = 5768 two hundred times
PRODUCT         1349712    = 5768 · 234
```

Now one step at a time:

Step one.

```
   323 ←——— CARRIED NUMBERS
  5768        Multiply (always from right to left)
×    4        8 × 4 = 32
 23072        6 × 4 = 24 + 3 = 27
              7 × 4 = 28 + 2 = 30
              5 × 4 = 20 + 3 = 23
```

Pay close attention to how your answers are written and to which numeral is carried.

Step two.

```
   2 2 2
   5768        Can you follow each step?
×    30        Where does this zero come from?
 173040        Remember the true statement "zero times any
               number equals zero."
```

Another way to solve this kind of problem is to write 30 in expanded form (3 × 10) and then multiply, first by ten, second by three. (Here is a rule that will help you. To multiply any number by ten put [add or annex] a zero on the right-hand side of the number.) Thus,

```
              2 2 2
5768 · 10 =  57680
            ×    3
            173040
```

Step three.

```
   1 1 1
   5768
×   200
 1153600
```

Check each step to be sure you understand.

Again this problem can be solved by writing 200 in expanded form (2 × 100) and multiplying. (The rule for multiplying by a hundred says to put two zeros on the right-hand side of the number.) So,

$$5768 \cdot 100 = \begin{array}{r} \overset{1\ 1\ 1}{576800} \\ \text{times}\underline{2} \\ 1153600 \end{array} \text{ and}$$

When you add the three partial products (23,072 + 173,040 + 1,153,600) you get the answer (product) 1,349,712.

Can you do the same problem by adding? Yes, you can. Use the rules for multiplying by 10, 100, 1000, and so forth. If you add 1 zero to multiply by 10, 2 zeros to multiply by 100, how many zeros would you add to multiply by 1000?

Multiply:
$$\begin{array}{r} 5768 \\ \times 234 \\ \hline 23072 \\ 173040 \\ 1153600 \\ \hline 1{,}349{,}712 \end{array}$$

By addition: carried numerals are circled

$5768 \cdot 4$

$(5768 \cdot 10) \cdot 3$
$57680 \cdot 3$
$5768 \cdot 30$

$(5768 \cdot 100) \cdot 2$
$576800 \cdot 2$

$$\begin{array}{r} ⑥⑤③ \\ 5768 \\ 5768 \\ 5768 \\ ⑤5768 \\ 57680 \\ 57680 \\ ③57680 \\ 576800 \\ 576800 \\ \hline 1{,}349{,}712 \end{array}$$

If you have an adding machine, this method for solving multiplication problems is relatively fast and easy. But, try to understand *why* you can use the adding machine to multiply.

You have probably noticed that different symbols have been used for the multiplication directive. Modern mathematics prefers to use the raised dot (·) to tell you to multiply. The more traditional math books use the "×" (called the times sign) to tell you to multiply. Whenever you see either the raised dot (·) or the times sign (×), you read or say "times." The raised dot is being given preference because of the possibility of confusion with the "x" in algebra. ("x" in algebra stands for an unknown quantity.) It is unlikely that the "×" will disappear from usage in the near future, so you should be familiar with, and able to use, both signs.

In algebra and modern mathematics there are other signs that say "multiply" too. You may see parentheses used this way: 4(7 + 3), which says 4 times 7 + 3, or 4 times 10 or 40; 4x in algebra and modern mathematics says four times x, where x can be any number. If x stands for 7 then 4x means 4(7) or 4 · 7 or 4 × 7 or 4 times 7 or 28.

While all the multiplication problems here have referred to the terms "multiplicand," "multiplier," and "product" in order to make the explanatory material more easily understood, there are other terms that can be used. Instead of multiplicand × multiplier = product, you can say factor × factor = product. The term "factor" is preferred in the modern mathematics.

HOW TO FACTOR

A <u>factor</u> is one of two or more numbers which when multiplied produces a product. Thus, factor times factor equals product, or 3 × 7 = 21, where 3 and 7 are factors and 21 is the product. Since 3 and 7 are prime numbers, they are called <u>prime factors.</u>

A number is said to be prime when it can be divided evenly by ONLY 1 and the number itself. Thus, 17 is prime because it can be divided evenly by ONLY 1 and 17. There are many prime numbers, some of which are listed in the table on page 16. Memorize the following prime numbers: 2, 3, 5, 7, 11, 13, 17, 19, 23, 29, 31, and 37. While there are many prime numbers, these can be most helpful to you.

Multiplication is the process of finding a product when two factors are known. The inverse operation of multiplication is division. The process of finding an unknown factor when one factor and the product are known is called division. In traditional division problems the factors will be named the divisor and the quotient, while the dividend corresponds to the product.

If a number can be divided evenly by more than two numbers, which is to say if a number can be divided by more numbers than 1 and the number itself, or if a number has more than two factors, it is called a <u>composite number.</u> All numbers that are not prime are composite. For example, 12 is a composite number. 12 has more than two factors. The factors of twelve are 1, 2, 3, 4, 6, 12. The factors produce the product 12 as follows:

$$1 \times 12 \quad\quad 12 \times 1$$
$$2 \times 6 \quad\quad 6 \times 2$$
$$3 \times 4 \quad\quad 4 \times 3$$

Complete factorization produces factors which are all prime numbers. Each composite number has only one set of prime factors. The prime factors of 12 are 2, 2, and 3. The product is gotten by multiplying 2 × 2 × 3, or said another way, $2^2 \times 3$. Note that all the factors are prime numbers.

The following method of stating the composite number as a product of prime numbers is called <u>branching</u> or setting up a "factor tree."

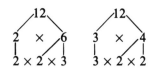

Another, and probably more efficient, method of complete factorization is to repeatedly divide the composite number by prime numbers. Start with 2 and move up the line until you get 1 as a quotient. For example:

```
2|12
 2|6
  3|3
    1
```

The prime factors of 12 are 2 × 2 × 3.

```
2|24
2|12
 2|6
 3|3
  1
```

The prime factors of 24 are 2 × 2 × 2 × 3.

```
3|27
 3|9
 3|3
  1
```

The prime factors of 27 are 3 × 3 × 3 or 3^3.

Sometimes though, it may be easier if you don't start with the lowest prime number. For example:

```
5|375
 5|75
 5|15
  3|3
   1
```

The prime factors of 375 are 3 × 5 × 5 × 5 or $3 × 5^3$.

```
3|9000
3|3000
2|1000
 2|500
 5|250
  5|50
  5|10
   2|2
    1
```

You do not have to follow a definite order, but always divide by prime numbers. Division must be even. The prime factors of 9000 are 2 × 2 × 2 × 3 × 3 × 5 × 5 × 5 or $2^3 × 3^2 × 5^3$.

Factorization can be very helpful in reducing fractions, finding the least common multiple, or the lowest common denominator. You can even do some division by factorization!

Reducing fractions by factorization:

$$\frac{25}{75} = \frac{\cancel{5} \times \cancel{5} \times 1}{3 \times \cancel{5} \times \cancel{5}} = \frac{1}{3}$$

Cancelling the 5's leaves the fraction reduced. Cancelling is actually the process of dividing both the numerator and the denominator by the same number. In cases like this, the quotient is 1.

$$\frac{375}{9000} = \frac{\cancel{3} \times \cancel{5} \times \cancel{5} \times \cancel{5}}{2 \times 2 \times 2 \times 3 \times \cancel{3} \times \cancel{5} \times \cancel{5} \times \cancel{5}} = \frac{1}{24}$$

One is always a factor if no other number is left.

After cancelling, multiply the remaining factors to get the product. Find the lowest common denominator of the following fractions.

$$\frac{1}{3}, \frac{1}{4}, \frac{1}{5}$$

What is really being sought is the lowest multiple of 3, 4, and 5 that is common to all of them. The multiples are really products and can be found by going through the multiplication tables for each number until you find a common product. Like this:

3 × 1 = 3	4 × 1 = 4	5 × 1 = 5
3 × 2 = 6	4 × 2 = 8	5 × 2 = 10
3 × 3 = 9	4 × 3 = [12]	5 × 3 = 15
3 × 4 = [12]	4 × 4 = 16	5 × 4 = △20
3 × 5 = (15)	4 × 5 = △20	5 × 5 = 25
3 × 6 = 18	4 × 6 = [24]	5 × 6 = 30
3 × 7 = 21	4 × 7 = 28	5 × 7 = 35
3 × 8 = [24]	4 × 8 = 32	5 × 8 = 40
3 × 9 = 27	4 × 9 = 36	5 × 9 = 45
3 × 10 = 30	4 × 10 = 40	5 × 10 = 50
3 × 11 = 33	4 × 11 = 44	5 × 11 = 55
3 × 12 = 36	4 × 12 = 48	5 × 12 = (60)
3 × 13 = 39	4 × 13 = 52	
3 × 14 = 42	4 × 14 = 56	
3 × 15 = 45	4 × 15 = (60)	
3 × 16 = 48		
3 × 17 = 51		
3 × 18 = 54		
3 × 19 = 57		
3 × 20 = (60)		

Twelve is a common multiple of 3 and 4. Fifteen is a common multiple of 3 and 5. Twenty is a common multiple of 4 and 5. But, it is not until we reach 60, that we find the first number that can be divided evenly by 3, 4, and 5. In this case, the lowest common multiple is the product of 3, 4, and 5. By multiplying all the denominators you can always get a common multiple, but it may not be the lowest common multiple and you may need to reduce your fraction before you have solved your problem. Remember, the common multiple produces a number that is divisible by (can be divided evenly by) all the denominators.

Let's look at another way to get the lowest common multiple. Use the fractions $\frac{1}{10}$, $\frac{1}{12}$, and $\frac{1}{16}$. If you multiply 10 × 12 × 16 your product is 1920 and that product is a common multiple of 10, 12, and 16. Is 1920 the least common multiple? Take each denominator and factor it completely.

10 = 2 × 5
12 = 2 × 2 × 3
16 = 2 × 2 × 2 × 2

Now take each individual prime number the number of times it appears most with any single denominator and solve for a product. For example, 2 is a factor of 10 once, of 12 twice, and of 16 four times. Therefore, write 2 as a factor four times. Three and five appear as factors only once and so only one of each will be used as a factor to get the least common multiple. Multiply all the factors and the product is the least common multiple.

L.C.M. = 2 × 2 × 2 × 2 × 3 × 5 = 240

So, 240 is the lowest common multiple, not 1920. You will find this method of solving for the least common multiple to be most helpful. Try to learn and understand it.

Now how can factoring help with division? There are times when it can't, but most of the time factoring can make division simpler. Take $32\overline{)7648}$ rewritten as a fraction, $\frac{7648}{32}$, and factor the numerator and the denominator.

$$
\begin{array}{rr}
2\underline{|7648} & 2\underline{|32} \\
2\underline{|3824} & 2\underline{|16} \\
2\underline{|1912} & 2\underline{|8} \\
2\underline{|956} & 2\underline{|4} \\
2\underline{|478} & 2\underline{|2} \\
239\underline{|239} & 1 \\
1 &
\end{array}
$$

$$\frac{7648}{32} = \frac{\not{2} \times \not{2} \times \not{2} \times \not{2} \times \not{2} \times 239}{\not{2} \times \not{2} \times \not{2} \times \not{2} \times \not{2} \times 1} = 239$$

You will find by referring to the prime number chart that 239 is a prime number. Since the prime factors of 7648 are $2^5 \times 239$ and the prime factors of 32 are 2^5, the 5 two's can be cancelled leaving $239 \div 1$, or 239, which is the quotient of the problem $7648 \div 32$. At first you may question whether or not this is an easier way to divide. This method does work with many problems and helps make math fun and interesting for many people. Try it and you might find a new challenge for yourself.

HOW TO SUBTRACT

Subtraction is an inverse operation of addition. The word "inverse" means the reverse method. When you go in reverse you go backwards. Pretty much the same thing happens with subtraction. Subtraction is related to addition as illustrated below.

a. 3 + 5 = 8
b. 3 + □ = 8
c. □ + 5 = 8
d. 8 − 3 = 5

In "supply the missing addend" problems you get the answers by the process of subtraction. You can solve the problem in example d by subtracting (taking away) the addend 3, as shown in example d.

Start with an array (group) of 8. Fence off 3. How many are left?

The same reasoning applies to example c.

Subtraction uses different designations and symbols. The top number represents the quantity from which another quantity will be taken and is called the *minuend*. The bottom number is the quantity to be subtracted and is called the *subtrahend*. The answer to a subtraction problem is called the *difference* or *remainder*. The sign for directing the subtraction operation is called the minus sign (−). For example:

```
                345  MINUEND
(MINUS SIGN)   −132  SUBTRAHEND
                213  REMAINDER OR DIFFERENCE
```

There is one more possibility of conflict in signs with subtraction. In foodservice, if you work in refrigeration, you may use, or see used, the minus sign to show the quantity of degrees below zero. So −12° is read "twelve degrees below zero." You may also run into the minus sign when you work with the number line. Some textbooks will show negative numbers using a raised dash to keep from confusing the negative sign with the minus sign (thus negative three is written ⁻3).

In the process of subtraction you use a "taking away" or "removing" of elements or quantities to find a remainder. If each digit in the minuend is larger than the corresponding digit directly below it in the subtrahend, simple subtraction can take place. For example:

```
 79865  MINUEND
−12543  SUBTRAHEND
 67322  REMAINDER
```

Begin with the one's column and move to the left.

Step one.   (one's column) Take 3 away from 5 leaving 2 remainder.

Step two.   (ten's column) Take 4 from 6 leaving 2 remainder.

Step three.   (hundred's column) Take 5 from 8 leaving 3 remainder.

Step four.   (thousand's column) Take 2 from 9 leaving 7 remainder.

Step five.   (ten thousand's column) Take 1 from 7 leaving 6 remainder.

You can now read your total remainder, which is your answer.

BORROWING: SUBTRACTION WITH REGROUPING

Learn these symbols. You will need to be able to read them in the discussion.

$>$ means is greater than.
$<$ means is less than.

If any digit in the minuend is smaller than the digit directly below it in the subtrahend, you cannot subtract until you have regrouped so a larger number appears on top. In the example below the 7 is greater than the 5 ($7>5$) so subtraction can't take place until the regrouping process has been completed.

$$\begin{array}{r} 35 \\ -27 \end{array}$$

Thirty-five is 3 tens and 5 ones but so is 2 tens and 15 ones. Borrowing is necessary when trying to explain the regrouping. The 3 in the ten's column means 30 (3×10). When added to the 5 ones, the sum is 35.

```
         ┌────── 3 TENS
         │  ┌─── 5 ONES
         ↓  ↓
30 + 5 = 35
20 + 15 = 35
```

The need is to get a number into the one's column from which 7 can be taken. Three tens and 5 ones can be regrouped in many ways so that subtraction can take place. (See below.)

$$\begin{array}{r} 27 + 8 = 35 \\ \text{MINUS } \underline{20 + 7 = 27} \\ 7 + 1 = 8 \end{array}$$

However, regrouping can be done more efficiently by regrouping in groups of ten. So, if 35 is regrouped into 2 tens and 15 ones, subtraction can readily take place. (See below.)

$$\begin{array}{r} (20 + 15) = 35 \\ -(20 + 7) = -27 \\ \hline 8 = 8 \end{array}$$

If, however, the regrouping can take place right in the problem, time can be saved. Here is where the term "borrowing" comes in. Follow this reasoning. Seven can't be subtracted from 5 so go to the ten's column and borrow 1 ten leaving 2 tens.

$$\begin{array}{r} 35 \\ -27 \end{array} \qquad \begin{array}{r} \overset{2}{\not{3}}5 \\ -27 \end{array}$$

Now take the 10 borrowed and add it to the 5 ones making 15. Then, subtract. (15 take away 7 leaves 8.)

$$\begin{array}{r} \overset{2}{\not{3}}15 \\ -2\ 7 \\ \hline 8 \end{array}$$

Solve this example. Then check these steps.

$$\begin{array}{r} 935 \\ -627 \end{array}$$

Step one. 7 > 5, so borrow 1 ten from the 3 in the ten's column. Regroup into 2 tens and 15 ones.

Step two. Take 7 from 15.

Step three. Take 2 from 2. (ten's column)

Step four. Take 6 from 9. (hundred's column)

$$\begin{array}{r} \overset{2}{9\not{3}}15 \\ -62\ 7 \\ \hline 8 \end{array} \qquad \begin{array}{r} \overset{2}{9\not{3}}15 \\ -62\ 7 \\ \hline 30\ 8 \end{array}$$

SUBTRACTION INVOLVING REGROUPING AND ZEROS

$$\begin{array}{r} 6005 \\ -2307 \end{array}$$

When there are no counting numbers {1, 2, 3, . . .} in columns in the minuend, you must continue across the columns, moving to the left until you find a column containing a counting number before you can borrow or regroup. In the example above, 7 > 5, so that regrouping is necessary before subtraction can be started.

In the example, the first counting number that can be used for regrouping is in the thousand's column. Regrouping is done one column at a time and from left to right. So start regrouping with the 6 in the thousand's column. Borrow 1, leaving 5. (You now have 5000 plus 1005, or 6005 regrouped.) There is a 10 in the hundred's column. Borrow 1 and regroup. There is a 10 in the ten's column. Borrow 1 ten and regroup.

$$\begin{array}{r} 6005 \\ -2307 \end{array} \qquad \begin{array}{r} \overset{5}{\not{6}}\ 10\ 0\ 5 \\ \\ \overset{5}{\not{6}}\ \overset{9}{\not{10}}\ 0\ 5 \\ \\ \overset{5}{\not{6}}\ \overset{9}{\not{10}}\ \overset{9}{\not{10}}\ 15 \end{array}$$

There is a 15 in the one's column now, and subtraction can begin.

```
    5 9 9
    6̸ 1̸0 1̸0 15
  − 2  3  0  7
    3  6  9  8
```

Take 7 from 15 leaving 8. (one's column)
Take 0 from 9 leaving 9. (ten's column)
Take 3 from 9 leaving 6. (hundred's column)
Take 2 from 5 leaving 3. (thousand's column)

HOW TO MAKE CHANGE

There may be a valid question about the placement of this topic under subtraction. You may want to make reference to it under some other topic. The process of making change is changing itself. The bigger foodservice/lodging concerns are now supplying clerks with cash registers capable of showing the exact amount of change to be given. However, once again, you should learn to give change to a customer completely on your own, just in case that should be necessary.

The first thing to keep in mind in giving change is that it is a subtraction process. For example, if a room costs $12.75 and the customer hands the clerk a $20 bill to make payment, how much change does the customer have coming? That is definitely a subtraction problem.

$$\begin{array}{r} 1\ 9\ 9 \\ \$\cancel{2}\cancel{0}.\cancel{0}^1 0 \\ -\ 1\ 2\ .\ 7\ 5 \\ \hline \$7\ .\ 2\ 5 \end{array}\ \begin{array}{l} \text{BILL PRESENTED} \\ \text{ROOM CHARGE} \\ \text{CHANGE RETURNED} \end{array}$$

There may be a better or easier way to make change. This is a counting or adding process. Take the same example, stated slightly differently. Charge for a room is $12.75. The customer hands the clerk a $20 bill. Count the change. Now, assume that the clerk has rung up the $12.75 in the cash register, the drawer is open. He takes the $20 bill and places it on the shelf just above the open drawer and, after checking the register display window, counts, saying, "$12.75, 13, 15 and five is 20. Thank you, sir."

What is he doing? He is counting the change as he gives it. He reads the $12.75 from the register display window. That's how much the customer owes. That much remains in the register. He hands the customer a quarter and says 13 ($12.75 + .25 = $13), hands him 2 one dollar bills and says 15 ($13 + $2 = $15), hands him a $5 bill and says 20 ($15 + $5 = $20). Has he given the customer $7.25? Has he done any subtraction?

Even if the clerk writes down a problem and subtracts, or if he uses a cash register that computes the change, the clerk would be wise to count the change in the manner just discussed for his own protection.

On the following page is a change-making chart which makes it easy for the student to count change and to become familiar with many different currency situations and how to handle them. You should practice with many situations requiring making change. Use the chart first. Then, if possible, get some actual experience using a cash register.

CHANGE SHEET

Sale	Amount Received	Amount of Sale	Amount of Change	Change Distribution								
				1¢	5¢	10¢	25¢	50¢	$1	$5	$10	$20

Note: The numbers under Change Distribution represent the value of coins or bills. Place the digit representing how many of each you are returning in the appropriate space. If the answer is none, do not use a zero but leave an empty space.

Permission is hereby granted to the teacher using this book to reproduce this page as needed.

HOW TO DIVIDE

Division is the process of finding how many times a quantity is contained in another quantity. Division is the inverse operation of multiplication.

In the section on "How to Multiply," you learned that terms were used to express multiplication problems as follows:

multiplicand × multiplier = product

or

factor × factor = product.

If division is the inverse of multiplication, then we solve for missing factors as follows:

factor × □ = product   product ÷ factor = □

where the sign (÷) is read "divided by" and □ is read "missing factor."

Signs used to denote the operation of division include:

÷ read "divided by."

$\frac{a}{b}$ read "a divided by b."

⌐ read "divided by" but is referred to by most people as the "gazinta" sign because most people when reading 6⌐12 would say "6 goes into 12."

Sign $\frac{a}{b}$ above indicates correctly that every fraction is a division problem.

The terms used in division are as follows:

                                           quotient
dividend ÷ divisor = quotient   divisor⌐dividend

$\frac{\text{dividend}}{\text{divisor}}$ = quotient   dividend/divisor = quotient

If the division is not exact (does not come out even), there is a remainder which follows the quotient. (See below.) The quotient is the answer to a division problem.

        quotient (and remainder)
divisor⌐dividend

The quotient and the divisor are related to the factors of a multiplication problem and the dividend is associated with the product. Using multiplication terms, division problems could be written

        factor
factor⌐product

if the division is exact, that is, does not have a remainder.

We have said that multiplication is an easy way to add. We have said too that division is the inverse operation of multiplication. If this is true, then division must be an easy way to subtract. In the examples that follow it can be seen that every division problem can be solved by using the subtraction operation. However, it is possible to make the subtraction process simpler. Study the following examples.

DIVISION BY SUBTRACTION

Even Division

How many times can 6 be taken from 48?

```
     8
  6)48      6 × 8 = 48        48
    48      48 ÷ 6 = 8       − 6  (1)
    ==                        42
                             − 6  (2)
     6                        36
  8)48      8 × 6 = 48       − 6  (3)
    48      48 ÷ 8 = 6        30
    ==                       − 6  (4)
                              24
                             − 6  (5)
                              18
                             − 6  (6)
                              12
                             − 6  (7)
                               6
                             − 6  (8)
                               0
```

Uneven division

How many times can 8 be taken from 51?

```
    8 r3         51
  6)51          − 8  (1)
    48           43
     ③         − 8  (2)
                 35
    6 r3       − 8  (3)
  8)51           27
    48         − 8  (4)
     ③           19
                − 8  (5)
                 11
                − 8  (6)
                  3  REMAINDER
```

There are many ways of estimating the number to subtract. One of the most popular ways to do this is as follows.

```
   47 | 3417      GROUPS    ⎧ This shows that ten groups of 47
      - 470        10       ⎨ can be subtracted. (47 × 10 = 470).
       2947                 ⎩
      -1410        30         47 × 30 = 1410
       1537
      -1410        30
        127
        - 94        2         47 × 2 = 94
         33        72
```

The quotient equals 72 remainder 33.
    Proof: Another proof:

```
     72            470
    ×47           1410
    ----          1410
    504             94
    288             33
   ----           ----
   3384           3417
    +33
   ----
   3417
```

All the numbers subtracted (really subtrahends) and the remainder add up to the sum 3417 which is equivalent to the dividend.

Another way of dividing by estimating and using subtraction has been successfully mastered by students of the authors' and is especially helpful when dividing by very large numbers. For example:

49347 | 987564123787

Step one involves continuous doubling of the divisor until the amount in one, two, four, and eight groups are known.

    49347 = 1 group
    98694 = 2 groups (The sum of 49347 and 49347.)
  197388 = 4 groups (The sum of 98694 and 98694.)
  394776 = 8 groups (The sum of 197388 and 197388.)

Now, pick the subtrahend and subtract. Place the partial quotient in the answer. (In this method always begin your subtraction on the left hand side of the dividend.)

```
                         20012647
           49347 | 987564123787
                  - 98694xxxxxxx
group chart        62412
49347 = 1         -49347
98694 = 2         130653
197388 = 8        -98694
394776 = 8        319597
      (4) =      -197388
  add            122209
      (2) =      -98694
                  235158
                 -197388
                  377707
      (4) =      -197388
                  180319
  add —(2) =     -98694
                   81625
      (1) =       -49347
                   32278
```

Since the divisor (49347) is contained in or can be subtracted from the first five numerals in the dividend, and since the five numerals in two groups can be subtracted from the first five numerals of the dividend, start with the number 98694 and subtract. The remainder is 62, which is smaller than the divisor (49347), so a 2 is placed in the quotient above the 6. The 4 is moved to the line for the new minuend making the number 624, but it is too small to subtract the divisor from so put a zero in the quotient and bring down the next numeral, 1, making the number 6241.

The divisor still can't be subtracted from this number so put another zero in the quotient and bring down the 2 making 62412. Subtract one group (the divisor) getting a remainder of 13065. Since only one group was subtracted place a 1 in the quotient, bring down the 3 and subtract. Observation of the group chart indicates that two groups can be subtracted. Subtract 98694, leaving a remainder of 31959. Since no further subtraction can take place at this point, bring down the 7 making 319597.

Observation of the group chart indicates that eight groups contain too large a number, but four groups can be subtracted. After subtracting four groups (197388) the remainder is 122209, so at least two more groups can be subtracted. After subtracting 98694 there is a remainder of 23515. No more groups of 49347 can be subtracted at this point, so the number of groups subtracted is totalled and the sum (6) is written in the quotient.

The next step proceeds as before—bring down the next number in the divisor (8) and subtract. Continue the process until no more subtraction can be completed and write the final remainder as a part of the quotient.

The quotient in this example is 20,012,647 r32278 where r is read "remainder." Please note that zero is used to hold the place twice when subtraction is impossible and that x is used to help keep track of the numerals brought down in the dividend so that the same number won't be used twice.

The whole point is that the process of division has many routes to the quotient. You should learn one process thoroughly so you feel quite confident that you can successfully solve any ordinary division problem. When you are comfortable doing division, study the many other processes for fun, enjoyment, and education.

FINDING AN AVERAGE CHECK

Knowing how to figure the average check is valuable knowledge in the foodservice/lodging industry. Finding an average involves addition and division usually. Suppose you wanted to find your average weekly math grade from five daily classes. To get the average, divide the sum of the grades by the number of grades. The quotient is your average grade. In the example, the average grade is 84.

```
    M =   75          84
    T =   90       5|420
    W =   85         40
    T =   90         20
    F =   80         20
  SUM =  420
```

In one hour Fay served fifteen customers lunch. She took in a total of $41.25. What was her average check for that hour?

```
        $2.75
15)$41.25
    30 xx
    112
    105
      75
      75
```

In Fay's case it isn't necessary to add all her checks since she knows the total already. She knows that each customer spent an average of $2.75 when she divides the total taken in by the number of customers served.

Figure a restaurant problem the same way. For example, Lyon's Restaurant serves 589 dinners in one evening for a total price of $2,609.27. How much was the average customer check? See if you get $4.43 for your answer.

HOW TO WORK WITH FRACTIONS

A fraction is a part of something, the word "fraction" itself meaning "to break." However, for use in mathematics consider a fraction to be a division problem, or a statement of a division problem which in reality represents a quotient. The modern mathematics presents fractions under the heading, rational numbers. A rational number is any number that can be expressed as a ratio or quotient of two whole numbers, provided the second number is not zero. A rational number is expressed with the terms a/b where a is a whole number and b is a counting or natural number. Since division by zero is meaningless, zero is excluded from the second term.

So a fraction is a division problem and names the quotient. However, once again the names of the parts are different. The parts of a fraction are called the terms of the fraction. The fraction is written $\frac{a}{b}$ where a and b are the terms of the fraction. The top number or term (a) is called the numerator; the bottom term (b) is called the denominator. If this same comparison is continued you have the following equivalents result.

$$\frac{a}{b} = \frac{\text{numerator}}{\text{denominator}} = \frac{\text{dividend}}{\text{divisor}} = \text{quotient}$$

A common fraction (sometimes called a simple fraction or a vulgar fraction) is a fraction whose numerator and denominator are both whole numbers or integers, except that the denominator cannot be a zero. If a fraction has a fraction for a numerator or denominator, it is called a complex fraction. A fraction is called proper when its numerator is smaller than its denominator, and is called improper if the numerator is larger than the denominator. If the numerator is larger than the denominator, the quotient is a mixed number or whole number.

The reciprocal of a given number is the factor which, when multiplied by the given number, produces a product of 1. Thus, the reciprocal of 2 is $\frac{1}{2}$; of 3 is $\frac{1}{3}$; of 8 is $\frac{1}{8}$. The reciprocal of a fraction is the fraction inverted. Thus, the reciprocal of $\frac{1}{2}$ is $\frac{2}{1}$; of $\frac{2}{3}$ is $\frac{3}{2}$; of $\frac{4}{5}$ is $\frac{5}{4}$. For example,

$$\frac{4}{5} \cdot \square = 1 \qquad \frac{4}{5} \cdot \frac{5}{4} = \frac{20}{20} = 1$$

so the reciprocal of $\frac{4}{5}$ is $\frac{5}{4}$ because $\frac{5}{4}$ is the number that gives the product 1 when multiplied by $\frac{4}{5}$.

Fractions that have the same denominators are called like fractions. If the denominators are not the same, the fractions are said to be unlike fractions. In some cases operations involving like fractions are simpler than with unlike fractions.

HOW TO ADD LIKE FRACTIONS

Remember: Like fractions have the same denominators. Examples of like fractions would include

$$\frac{1}{6}, \frac{2}{6}, \frac{3}{6}, \frac{4}{6}, \frac{5}{6}, \frac{6}{6}, \frac{7}{6}, \frac{8}{6} \ldots$$

Rule: To add like fractions add the numerators and write their sum over the same denominator. For example:

$$\frac{1}{6} + \frac{2}{6} = \frac{3}{6}$$

Procedure: Add $1 + 2 = 3$; write the 3 over the like denominator (6) and secure the sum $\frac{3}{6}$.

Now for a further explanation of the terms. Earlier it was said that a fraction is a part of something. The fraction names the part of a whole thing and tells how many of those parts are used in a particular fraction. So the fraction $\frac{1}{6}$ delivers two messages; one message (the denominator) tells that the whole is divided into six equal parts, the other message (the numerator) tells that one of those six parts is being used. Once again, the terms of the fraction are the numerator and the denominator. The denominator tells how many equal parts are in a whole; the numerator tells how many of those parts are being used.

The fraction $\frac{1}{6}$ tells that you are using one out of six parts, $\frac{2}{6}$ says two out of six, $\frac{6}{6}$ says 6 out of 6, or all the parts of a whole. So $\frac{6}{6}$ equals one whole.

Suppose you were asked how many dimes you had. In one pocket you find 3 dimes, in another 4 dimes. How many dimes do you have? You have 7 dimes. The word "dime" names the part you are adding. Try it like a fraction.

$$\frac{3}{\text{dimes}} + \frac{4}{\text{dimes}} = \frac{7}{\text{dimes}}$$

Now, since a dime is a tenth of a dollar, the denominator above (dimes) could be changed to tenths and still have the same meaning. Written in arithmetic form the problem becomes

$$\frac{3}{10} + \frac{4}{10} = \frac{7}{10}$$

In this form the denominator does not necessarily represent money, but it can. One other thing to note in this example—the dime here represents a fraction, a tenth of a dollar, but the dime could be considered a whole too. (What coin represents a tenth of a dime?) This will be true many times in the everyday problems you run into in the foodservice/lodging industry.

HOW TO ADD UNLIKE FRACTIONS

In order to add fractions that have unlike denominators, the fractions must first be changed so the denominators are the same.

In the foodservice business you might be instructed to put $\frac{1}{2}$ cup of sugar, $\frac{2}{3}$ cup of rice flour, and $\frac{3}{4}$ cup of wheat flour into a bowl. How many cups of material are in the bowl? In order to solve the problem arithmetically you must add $\frac{1}{2}$, $\frac{2}{3}$, and $\frac{3}{4}$. You cannot add fractions if the denominators are not the same.

FRACTIONS ON THE NUMBER LINE

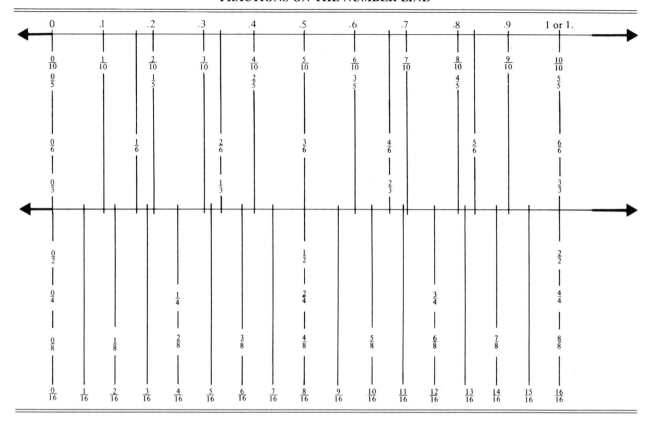

At this point reference to the Equivalent Fractions table on page 15 might be most helpful. Refer to the fractions that have 2, 3, and 4 as denominators at the beginning of lines. Next read across each line until the lowest common denominator is found. It is 12. Since each line of fractions begins with a numerator of 1, care must be taken in writing the equivalents. $\frac{1}{2} = \frac{6}{12}$; $\frac{1}{3} = \frac{4}{12}$; $\frac{1}{4} = \frac{3}{12}$ according to the chart. The problem needs equivalents for $\frac{1}{2}$, $\frac{2}{3}$, and $\frac{3}{4}$. Therefore, the numerator of the equivalent for $\frac{1}{3}$ must be multiplied by 2 and for $\frac{3}{4}$ must be multiplied by 3. This process gives equivalents as follows:

$$\frac{1}{2} = \frac{6}{12}$$
$$\frac{2}{3} = \frac{8}{12}$$
$$\frac{3}{4} = \frac{9}{12}$$

Add the numerators either vertically

$$\frac{6}{12}$$
$$\frac{8}{12}$$
$$\frac{9}{12}$$
$$\overline{\frac{23}{12}} = 1\frac{11}{12} \text{ cups}$$

or horizontally.

$$\frac{6}{12} + \frac{8}{12} + \frac{9}{12} = \frac{23}{12} \text{ cups}$$

Notice that the answer is $\frac{23}{12}$ cups or $1\frac{11}{12}$ cups. $1\frac{11}{12}$ cups is the preferred answer because the answer is easily understood. In mathematics it is said that $1\frac{11}{12}$ is the simpler answer. $\frac{23}{12}$ is an improper fraction, whereas $1\frac{11}{12}$ is a mixed number.

Now, let's review: To add like fractions, add the numerators, and write the sum over the like denominator. To add unlike fractions, first change the fractions to equivalent fractions with the same denominators, then add the numerators, and write the sum over the common denominator. Express the answer in simplest form.

HOW TO FIND A COMMON DENOMINATOR

Suppose you don't have an equivalent fractions table, or the table is not detailed enough to permit finding a denominator that is common to all fractions, what do you do? One method is to put your ability to factor numbers to work for you.

Find the common denominator of $\frac{1}{12}$, $\frac{1}{15}$, and $\frac{1}{35}$. Factor the denominators first

$12 = 2 \cdot 2 \cdot 3$
$15 = 3 \cdot 5$
$35 = 5 \cdot 7$

Write the prime factors to get the lowest common multiple. L.C.M. = $2 \cdot 2 \cdot 3 \cdot 5 \cdot 7$. Multiply. L.C.M. = 420.

The product is the lowest common multiple, or the lowest common denominator. Now the problem changes from $\frac{1}{12}$, $\frac{1}{15}$, and $\frac{1}{35}$ to $\frac{35}{420}$, $\frac{28}{420}$, and $\frac{12}{420}$. Now the fractions can be added.

If <u>all</u> the denominators are prime numbers as in $\frac{1}{2}$, $\frac{1}{3}$, and $\frac{1}{5}$, the lowest common denominator is the product of the denominators. So the L.C.D. in this case is $2 \cdot 3 \cdot 5$, or 30. At other times you can just look at the denominators and pick out the common denominator. Take $\frac{1}{2}$, $\frac{1}{4}$, and $\frac{1}{8}$. It is readily observed that both 2 and 4 will divide evenly into 8, and so you immediately know that 8 is the lowest common denominator.

HOW TO CHANGE FRACTIONS

Many times it is more convenient to work with fractions in a different form. Most of the time answers to problems involving fractions are expected in simplest terms. In order to operate with complete ease on fraction problems, you need to know how to reduce fractions and how to raise fractions to higher terms.

The Equivalent Fractions table (page 15) will help in many cases. Start with the first fraction on the left-hand side of the table in any row. The other fractions in that same row are multiples of the first fraction. However, if you start with any fraction in a row except the first fraction on the left-hand side, you will notice that the fraction can always be reduced to the first fraction in that row.

A fraction is said to be in lowest terms (that is, reduced or simplified) when those terms can be divided evenly by 1 only. So $\frac{1}{2}$ is in lowest terms because the 1 and the 2 cannot be divided by any other number but 1. The same is true for $\frac{3}{4}$, $\frac{5}{6}$, and so on. Reducing then becomes a matter of factoring. Remember, the terms of a fraction are the numerator and the denominator. When a fraction is in lowest terms, its numerator and denominator have only one common factor, the number 1. To reduce $\frac{9}{12}$ to lowest terms factor the numerator and denominator to prime factors.

$$\frac{9}{12} = \frac{3 \times 3}{2 \times 2 \times 3}$$

Then cancel the common factors and write the answer in reduced terms. To cancel means to remove all the factors equivalent to 1. This is rewritten:

$$\frac{9}{12} = \frac{1}{2} \times \frac{3}{2} \times \frac{3}{3}$$

Since $\frac{3}{3} = 1$, we cancel the 3's out.

$$\frac{1}{2} \times \frac{3}{2} \times \frac{\cancel{3}}{\cancel{3}} \quad \text{leaving} \quad \frac{1}{2} \times \frac{3}{2} = \frac{1 \times 3}{2 \times 2} = \frac{3}{4}$$

By factoring we find that the terms 9 and 12 can be divided evenly by 3. Carrying out the division gives the same lowest term fraction.

$$\frac{9 \div 3}{12 \div 3} = \frac{3}{4}$$

Sometimes the common divisor can be chosen by observation. For example, you might be able to look at the fraction $\frac{9}{12}$ and think both 9 and 12 are multiples of 3, so 3 will divide both evenly. If both terms are even numbers you can immediately cut the terms in half. Once in a while, by repeatedly cutting terms in half, you can reach lowest terms as in

$$\frac{24}{32} = \frac{12}{16} = \frac{6}{8} = \frac{3}{4}.$$

Use the method you find easiest. However, it will probably be best to master the factoring method.

HOW TO SUBTRACT LIKE FRACTIONS

To subtract like fractions, subtract the numerators and write the remainder over the common denominator. Reduce the answer to lowest terms if necessary. For example:

$$\frac{5}{6} - \frac{3}{6} = \frac{2}{6} = \frac{1}{3}$$

or

$$\begin{array}{r} \frac{5}{6} \\ -\frac{3}{6} \\ \hline \frac{2}{6} = \frac{1 \cdot 2}{3 \cdot 2} = \frac{1}{3} \end{array}$$

HOW TO SUBTRACT UNLIKE FRACTIONS

To subtract unlike fractions, first change the fractions so they have common denominators, then proceed as with like fractions.

$$\frac{2}{3} - \frac{1}{4} =$$

$$\frac{8}{12} - \frac{3}{12} = \frac{5}{12}$$

$$\frac{26}{39} - \frac{3}{33} =$$

$$\frac{26 \times 11}{39 \times 11} = \frac{286}{429}$$

$$\frac{3 \times 13}{33 \times 13} = -\frac{39}{429}$$

$$\frac{247}{429}$$

FACTOR
$39 = 3 \cdot 13$
$33 = 3 \cdot 11$
L.C.D. $= 3 \cdot 11 \cdot 13 = 429$

$429 \div 39 = 11$
$429 \div 33 = 13$

How do you go about reducing $\frac{247}{429}$ s?

Reference to the Prime Number chart shows neither 247 nor 429 to be prime.

$247 \div 3$ is uneven.
$429 \div 3$ is even.
$247 \div 11$ is uneven.
$429 \div 11$ is even.
$247 \div 13$ is even.
$429 \div 13$ is even.

You can start dividing both 247 and 429 by the same prime number until you discover one or more common divisors. Or, since you know the prime factors of 429 are 3, 11, and 13, you can begin dividing by them to see if any factors are common. (This is the recommended way.) Division indicates that 13 is a common factor of 247 and 429. Easier still, refer to the multiplication tables and find that $13 \cdot 19 = 247$. Since both 13 and 19 are prime, those are the factors of 247.

$$\frac{247}{429} = \frac{13 \times 19}{3 \times 11 \times 13} = \frac{19 \times 13}{3 \times 11 \times 13} = \frac{19}{33} \times \frac{13}{13} = \frac{19}{33} \times 1 = \frac{19}{33}$$

$$\frac{26}{39} = \frac{286}{429}$$
$$-\frac{3}{33} = \frac{39}{429}$$
$$\frac{247}{429} = \frac{19}{33}$$

Before leaving this section, look again at the part of the problem shown below. We determined that 429 was the common denominator and that 39 and 11 were the factors needed in solving the problem.

$$\frac{26}{39} \times \frac{11}{11} = \frac{286}{429}$$
$$-\frac{3}{33} \times \frac{13}{13} = \frac{39}{429}$$

We multiplied 39 by 11 to get 429, but why did we multiply 26 by 11? Remember the rule that when a number is multiplied by 1 the product is the number itself? The number 1 is the identity number in multiplication. Is $\frac{11}{11}$ the same as 1, another name for 1?

Since multiplication of a number by 1 does not change the value of the number, when we multiplied the fractions $\frac{26}{39}$ and $\frac{3}{33}$ by fractions that were equivalent to 1 ($\frac{11}{11}$ and $\frac{13}{13}$), we did not change their value. We multiplied 26 × 11 so the product (286) would have the same relationship to 429 that 26 has to 39. By doing this both fractions have the same value and are equivalent.

After reading this explanation you may understand why some mathematicians think that the first process that should be taught when beginning fractions is multiplication. Here knowledge of the multiplication process is needed to add and subtract unlike fractions.

HOW TO CHANGE MIXED NUMBERS FOR ADDITION AND SUBTRACTION OF FRACTIONS

A mixed number is another form of an improper fraction. Sometimes it is easier to work with mixed numbers in the improper fraction form. Therefore, it is necessary to be able to change quickly from one form to the other.

To change a mixed number to an improper fraction, change the whole number to fractional form having the same denominator as the fraction and add the fraction. For example:

$$1\tfrac{1}{2} = 1 + \tfrac{1}{2} = \tfrac{2}{2} + \tfrac{1}{2} = \tfrac{3}{2}$$

$$2\tfrac{3}{5} = 2 + \tfrac{3}{5} = 1 + 1 + \tfrac{3}{5} = \tfrac{5}{5} + \tfrac{5}{5} + \tfrac{3}{5} = \tfrac{10}{5} + \tfrac{3}{5} = \tfrac{13}{5}$$

A shortcut

$$2\tfrac{3}{5} = \frac{5 \times 2 + 3}{5} = \frac{10 + 3}{5} = \frac{13}{5}$$

Note the manner in which the numerals are written in the denominator and write out every mixed number the same way when first learning this method. Later you will be able to do this (follow the arrow):

$$4\tfrac{5}{7} = \tfrac{33}{7}$$

The arrows tell you to start with 7, multiply by 4, and add 5 (7 × 4 = 28 + 5 = 33). Then write the answer over the denominator.

To change an improper fraction to a mixed number divide the denominator into the numerator and write the quotient as a whole number and a proper fraction. For example:

$$\tfrac{3}{2} = 3 \div 2 = 1\tfrac{1}{2} \qquad\qquad \tfrac{33}{7} = 33 \div 7 = 4\tfrac{5}{7}$$

$$\tfrac{3}{2} = 2\overline{)3}\;\;\tfrac{1\tfrac{1}{2}}{} \qquad\qquad \tfrac{33}{7} = 7\overline{)33}\;\;\tfrac{4\tfrac{5}{7}}{28}$$

$$\tfrac{3}{2} = \tfrac{1}{2} + \tfrac{1}{2} + \tfrac{1}{2} = \tfrac{2}{2} + \tfrac{1}{2} = 1 + \tfrac{1}{2} = 1\tfrac{1}{2}$$

To add mixed numbers or mixed numbers and fractions:

$1\frac{1}{2} + 1\frac{1}{2} =$

$\frac{3}{2} + \frac{3}{2} = \frac{6}{2} = 3$

$$\begin{array}{r} 1\frac{1}{2} \\ +1\frac{1}{2} \\ \hline 2\frac{2}{2} = 1 \\ +1 \\ \hline 3 \end{array}$$

$1\frac{1}{2} + \frac{2}{3} + \frac{3}{4} = \frac{3}{2} + \frac{2}{3} + \frac{3}{4} = \frac{18}{12} + \frac{8}{12} + \frac{9}{12} = \frac{35}{12} = 2\frac{11}{12}$

HOW TO OPERATE WITH DECIMALS

The term "decimals" is used to refer to decimal fractions, a term that is no longer used widely. Decimal fractions are considered another form of rational numbers. Here rational numbers are referred to as decimals because that is the term most often used in the industry, and many of the people you will work with have never used the term "rational numbers." Rational numbers are defined as all the numbers that can be represented as the quotient of two or more whole numbers, except that the denominator cannot be zero.

Decimal fractions refer to those fractions which are expressed with denominators of ten or powers of ten. Reference to the section on place value will demonstrate how the use of the decimal point and the digits in columns states a fraction in decimal form. On the number line, we learn that numbers go in both directions from zero. In place value the units (or one's) column is the first column to the left of the decimal point. As the columns move left each column is an increasing multiple of ten. Thus the columns moving left are the

THOUSANDS	HUNDREDS	TENS	ONES
1000	100	10	1

Notice that as you move to the right each column is one-tenth of the immediate left-hand column. What happens when you take one-tenth of one? The product, of course, is $\frac{1}{10}$. As we move to the right of the decimal point this is how each column gets its name and value.

$$1000 \quad 100 \quad 10 \quad 1 \cdot \frac{1}{10} \quad \frac{1}{100} \quad \frac{1}{1000} \quad \frac{1}{10000}$$

Sometimes you will see the columns referred to this way

$$10^3 \quad 10^2 \quad 10^1 \quad 10^0 \cdot 10^{-1} \quad 10^{-2} \quad 10^{-3} \quad 10^{-4}$$

The symbol 10^{-1} is just another way of writing $\frac{1}{10}$; 10^{-2}, of writing $\frac{1}{100}$; 10^{-3}, of writing $\frac{1}{1000}$; and so on. The negative sign tells you that the power of ten is a denominator with a numerator of one. So the first column to the right of the decimal point is named tenths, the second hundredths, the third thousandths, and so forth.

CHANGING FRACTIONS TO DECIMALS

To express proper fractions in decimal form, divide the numerator by the denominator.

$$\frac{2}{5} = 2 \div 5 = 5\overline{)2.0} \quad \text{Proof: } .4 = \frac{4}{10} = \frac{2}{5}$$
$$\phantom{\frac{2}{5} = 2 \div 5 = 5\overline{)}}\underline{2\ 0}$$

$$\frac{3}{8} = 8\overline{)3.000}^{.375}$$
$$\phantom{\frac{3}{8} = 8\overline{)}}\underline{2\ 4xx}$$
$$\phantom{\frac{3}{8} = 8\overline{)3.00}}60$$
$$\phantom{\frac{3}{8} = 8\overline{)3.00}}\underline{56}$$
$$\phantom{\frac{3}{8} = 8\overline{)3.00}}40$$
$$\phantom{\frac{3}{8} = 8\overline{)3.00}}\underline{\underline{40}}$$

Proof: $.375 = \frac{375}{1000} = \frac{3 \cdot 5 \cdot 5 \cdot 5}{2 \cdot 2 \cdot 2 \cdot 5 \cdot 5 \cdot 5} = \frac{3}{2 \cdot 2 \cdot 2} = \frac{3}{8}$

You may find it helpful to read the following paragraphs through division by decimals and then come back and reread this section on changing fractions to decimals.

HOW TO ADD OR SUBTRACT DECIMALS

To add or subtract decimal fractions be sure all decimal points are in the proper column and perform the operations as with whole numbers. This point is important. It does cause trouble. Keep all whole numbers in their proper columns. Keep all decimal fractions in their proper columns. The decimal point separates whole numbers from fractions. Therefore, the decimal points must be directly above and below each other.

```
                                      0        5
  .3      .6      .3      .9     1.7   1̸.¹7    6̸.¹4
 +.4     +.7      .4     -.3    - .5   - .9   - 1. 7
 ___     ___     ___     ___    ____   ____   _____
  .7     1.3     1.5      .6    1.2     .8    4. 7
                 6.0
                 ___
                 8.2
```

Watch out for whole numbers.

```
                            2 9
3 + .06 =  3.00   3 - .06 = 3̸.0̸¹0
          + .06            - .0 6
          _____            _____
           3.06             2.9 4
```

Subtract 56 cents from five dollars.

```
  4 9
 $5̸.0̸¹0
 - .5 6
 _____
 $4.4 4
```

Add: .08, .986, 1.2, 101.09, 3.67

```
      3
    .08
   2.986
   1.2
 101.09
   3.67
 _____
 107.026
```

The decimal points make the column that keeps all the digits in their proper places.

Be careful carrying numbers. In adding .9 to .4 you get a sum which is a mixed decimal (a mixed fraction in decimal form) or a whole number and a decimal. The whole number and the decimal are separated by a decimal point.

$$\begin{array}{r} .9 \\ \underline{.4} \\ 1.3 \end{array}$$

in another way:

$$\frac{9}{10} + \frac{4}{10} = \frac{13}{10} = 1\frac{3}{10} = 1.3$$

Remember: Borrowing is the same as with whole numbers.

HOW TO MULTIPLY DECIMALS

To multiply decimals proceed in the same manner as with whole numbers to get a product; then locate the decimal point so the product has the same number of fractional places as the two factors.

.3 ONE FRACTIONAL PLACE PLUS
×.4 ONE FRACTIONAL PLACE EQUALS
.12 TWO FRACTIONAL PLACES.

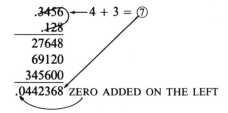

3.23 ← 2 + 1 = ③ FRACTIONAL PLACES
×1.5
1615
323
4.845

If you don't have enough digits in your product, annex (add, put on) zeros to complete the decimal.

.3456 ← 4 + 3 = ⑦
.128
27648
69120
345600
.0442368 ZERO ADDED ON THE LEFT

After finishing the multiplication, there were only six digits in the product. Seven places were needed so the zero was annexed <u>to the left</u> before placing the decimal point. Since all the fractional places are on the right-hand side of the decimal point, all you need to do is to count the places to the right of the decimal in both factors and then count off the places moving from right to left in the product.

HOW TO DIVIDE DECIMALS

To divide using decimals, proceed as if the numbers were whole numbers and locate the decimal point as follows:
    If dividing by a whole number, locate the decimal point directly above the decimal point in the dividend in the division box.

$$\begin{array}{r} .04 \\ 6\overline{\smash{)}.24} \\ \underline{24} \end{array}$$

Annex zeros as needed to hold the place.

```
      1.21          .34          .0005        .6         .625
    3│3.63       11│3.74       5│.0025      5│3.0      8│5.000
      3 xx          3 3x          25           3 0        4 8xx
      6             44            ══           ══          20
      6             44                                     16
      ─             ══                                     40
      3                                                    40
      3                                                    ══
      ══
```

If dividing a whole number or mixed decimal by a mixed decimal or a decimal fraction, change the divisor and the dividend so that the divisor is a whole number and the dividend is in the same relationship to the divisor as before by multiplying both by the same number. Annex zeros as needed.

$$.25\overline{)1.} = .25\overline{)1.00} = 25.\overline{)100.}$$

ANNEX ZEROS   MULTIPLY BY 100

Written as a fraction:

$$\frac{1}{.25} = \frac{1 \times 100}{.25 \times 100} = \frac{100}{25}$$

The quotient obviously is 4.

In this type of problem you will be multiplying both the divisor and dividend by multiples of ten. Remember you can multiply a number by 10 by adding a zero, by 100 by adding two zeros, etc. ($35 \times 10 = 350$; $67 \times 100 = 6700$). If you have a decimal fraction or a mixed decimal, it is easy to multiply and divide by powers of 10. To multiply by 10, move the decimal point one place to the <u>right</u>, to multiply by 100 move two places to the <u>right</u>, and so forth. To divide follow the same procedure to the <u>left</u>.

In division by decimal fractions or mixed decimals, this shortcut comes in very handy. Simply move the decimal point in the divisor so that the divisor is a whole number. Move the decimal point in the dividend to the right the same number of places and mark the place with a caret (∧). The caret is a symbol which indicates that something goes in its place or that something is to be inserted. In the quotient insert the decimal point directly above the caret, ∧.

Step one: Move divisor decimal point to make a whole number.

Step two: Move dividend decimal point the same number of places marking the place with the caret.

Step three: Divide and place the decimal point.

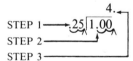

```
        .9
   1.25│1.125
        1 125
```

Divide 72 by .036.

```
         2000.
    .036│72.000
```

63

Another example:

$$.125\overline{)\,.437_\wedge 5\,}^{3.5}$$
$$\phantom{.125\overline{)}}\underline{375x}$$
$$\phantom{.125\overline{)}\,}62\ 5$$
$$\phantom{.125\overline{)}\,}\underline{62\ 5}$$

If you would rather, put a caret here too.

$$.33\overline{)\,.36_\wedge 3\,}^{1.1}$$
$$\phantom{.33\overline{)}}\underline{33}$$
$$\phantom{.33\overline{)}\,}3\ 3$$
$$\phantom{.33\overline{)}\,}\underline{3\ 3}$$

To round off use the digit in the first column to the right of the desired decimal place. This method of locating decimal points will be useful in most decimal problems. But what about remainders? Some problems just do not come out evenly. When they don't, either end the quotient with a fraction or round it off. Rounding off either raises the last digit needed by one or keeps it as is. If the fraction is equivalent to one-half or more, the quotient ends in the next highest digit. Examples.

$$3\overline{)1\,} = 3\overline{)1.00\,}^{.33\frac{1}{3}}$$
$$\phantom{3\overline{)1.0}}\underline{9x}$$
$$\phantom{3\overline{)1.}}10$$
$$\phantom{3\overline{)1.}}\underline{9}$$
$$\phantom{3\overline{)1.}}\tfrac{1}{3}$$

Decimals like this usually end in a fraction.

$$3\overline{)2.00\,}^{.66\frac{2}{3}}$$
$$\phantom{3\overline{)2}}\underline{1\ 8x}$$
$$\phantom{3\overline{)2.}}20$$
$$\phantom{3\overline{)2.}}\underline{18}$$
$$\phantom{3\overline{)2.}}\tfrac{2}{3}$$

But sometimes you can draw a line over the last digit and give the correct answer. The line over the 6 means that the 6 repeats forever.

$$3\overline{)2.000\,}^{.66\overline{6}}$$
$$\phantom{3\overline{)2.}}\underline{1\ 8xx}$$
$$\phantom{3\overline{)2.}}20$$
$$\phantom{3\overline{)2.}}\underline{18}$$
$$\phantom{3\overline{)2.}}20$$
$$\phantom{3\overline{)2.}}\underline{\underline{20}}$$

If we used this problem and gave the answer to the nearest hundredth, the quotient would be .67. Since the fraction is greater than $\frac{1}{2}$, use the next higher digit. ($\frac{2}{3} > \frac{1}{2}$ so use .67 instead of .66)

$$3\overline{)2.00\,}^{.66\frac{2}{3}\ =\ .67}$$

But in this problem, the fraction is less than $\frac{1}{2}$ so the digit does not change. ($\frac{1}{3} < \frac{1}{2}$ so use .33)

$$\begin{array}{r} .33\frac{1}{3} = .33 \\ 3\overline{)1.00} \\ \underline{9x} \\ 10 \\ \underline{9} \\ \frac{1}{3} \end{array}$$

Express $\frac{11}{19}$ as a decimal fraction correct to the nearest thousandth.

$\frac{11}{19} = .579$

There are two ways to get the answer.

$$\begin{array}{r} .578\frac{18}{19} = .579 \\ 19\overline{)11.000} \\ \underline{9\ 5xx} \\ 1\ 50 \\ \underline{1\ 33} \\ 170 \\ \underline{152} \\ \frac{18}{19} \end{array}$$

The 8 is raised to a 9 because $\frac{18}{19} > \frac{1}{2}$.

In the example below the 8 is raised to a 9 because the fourth decimal place is more than 5. If the next digit to the right of the desired decimal place is 5 or more, then the fraction is equal to or greater than $\frac{1}{2}$ and the desired quotient is expressed in the higher term.

$$\begin{array}{r} .5789 = .579 \\ 19\overline{)11.0000} \\ \underline{9\ 5xxx} \\ 1\ 50 \\ \underline{1\ 33} \\ 170 \\ \underline{152} \\ 180 \\ \underline{171} \end{array}$$

If the digit is less than 5, keep the digits as calculated. For example, express $\frac{1}{3}$ as a decimal fraction correct to the nearest hundredth. Since the digit in the thousandths place is less than 5, the digits remain unchanged and the quotient is expressed, correct to the nearest hundredth, as .33.

$$\begin{array}{r} .333 = .33 \\ 3\overline{)1.000} \\ \underline{9xx} \\ 10 \\ \underline{9} \\ 10 \\ \underline{9} \end{array}$$

OPERATING WITH WHOLE NUMBERS, MIXED NUMBERS, AND FRACTIONS

Addition

If whole numbers and fractions are to be added, just combine them. Examples.

$$2 + \frac{2}{5} = 2\frac{2}{5} \qquad \begin{array}{r} 2 \\ + \frac{2}{5} \\ \hline 2\frac{2}{5} \end{array}$$

The whole number represents 2 and no fifths and could be written this way.

$$\begin{array}{r} 2\frac{0}{5} \\ + \frac{2}{5} \\ \hline 2\frac{2}{5} \end{array}$$

However, why waste time writing $\frac{0}{5}$ if you understand the problem? You will usually find the whole number alone, i.e., without a fraction with a zero for the numerator.

Subtraction

To subtract when mixed numbers are involved follow the examples below.

$$1\frac{1}{2} - \frac{2}{3} =$$

Change mixed number to an improper fraction.

$$\frac{3}{2} - \frac{2}{3} =$$

Change to like fractions. Subtract.

$$\frac{9}{6} - \frac{4}{6} = \frac{5}{6}$$

Vertical method is almost the same as the horizontal.

$$\begin{array}{r} 1\frac{1}{2} = 1\frac{3}{6} = \frac{9}{6} \\ -\frac{2}{3} = -\frac{4}{6} = -\frac{4}{6} \\ \hline \frac{5}{6} \end{array}$$

Below, borrow 1 from the 3. Regroup, change, and subtract. Remember to subtract the whole numbers.

$$\begin{array}{r} \overset{2}{\cancel{3}}\frac{1}{2} = 1\frac{3}{6} = \frac{9}{6} \\ -\frac{2}{3} = -\frac{4}{6} = -\frac{4}{6} \\ \hline 2 \qquad\qquad \frac{5}{6} = 2\frac{5}{6} \end{array}$$

No borrowing or regrouping but a common denominator is necessary.

$$5\tfrac{5}{6} = \tfrac{10}{12}$$
$$-1\tfrac{1}{4} = \tfrac{3}{12}$$
$$\overline{\phantom{-1\tfrac{1}{4}=}}$$
$$4\tfrac{7}{12} = 4\tfrac{7}{12}$$

The answer is the same this way too.

$$5\tfrac{5}{6} - 1\tfrac{1}{4} = \tfrac{35}{6} - \tfrac{5}{4} = \tfrac{70}{12} - \tfrac{15}{12} = \tfrac{55}{12} = 4\tfrac{7}{12}$$

Subtracting fractions from whole numbers often means trouble for those who don't understand the process of changing a borrowed whole to a fraction. How do you regroup a whole into a fraction? When a number is borrowed in a subtraction problem, the borrowed number is a unit, one. The number one can always be regrouped into a fraction as long as the numerator and denominator are the same. For example:

$$1 = 1\tfrac{0}{4} = \tfrac{4}{4}$$
$$-\tfrac{3}{4} = -\tfrac{3}{4} = -\tfrac{3}{4}$$
$$\overline{\tfrac{1}{4}}$$

$$1 - \tfrac{3}{4} = \tfrac{4}{4} - \tfrac{3}{4} = \tfrac{1}{4}$$

Another example:

$$3 - \tfrac{3}{8} =$$

Regroup 3 as 2 + 1.

$$(2 + 1) - \tfrac{3}{8} =$$

Change 1 to $\tfrac{8}{8}$

$$\left(2 + \tfrac{8}{8}\right) - \tfrac{3}{8} =$$

Subtract the fractions

$$2 + \left(\tfrac{8}{8} - \tfrac{3}{8}\right) =$$

Combine for the answer.

$$2 + \tfrac{5}{8} = 2\tfrac{5}{8}$$

What about larger numbers? No matter how large the numbers, proceed in the same way.

$$123\tfrac{1}{3}$$
$$-89\tfrac{5}{8}$$
$$\overline{\phantom{-89\tfrac{5}{8}}}$$

Find the common denominator. Note that 15 cannot be subtracted from 8.

$$123\tfrac{1}{3} = \tfrac{8}{24}$$
$$-89\tfrac{5}{8} = \tfrac{15}{24}$$

Regroup the 123. (borrow)

$\quad 122 + 1$

Associate the 1 with the fraction.

$\quad 122 + 1\frac{8}{24}$

Change the mixed number to an improper fraction.

$\quad 1 \,{}^+_\times \frac{8}{24} = \frac{32}{24}$

The rewritten problem.

$$\begin{array}{r} 12\overset{2}{\cancel{3}}\frac{1}{3} = 1\frac{8}{24} = \frac{32}{24} \\ -89\frac{5}{8} = \frac{15}{24} = \frac{15}{24} \\ \hline \end{array}$$

Subtract the fractions.

$$\begin{array}{r} \frac{32}{24} \\ -\frac{15}{24} \\ \hline \frac{17}{24} \end{array}$$

Subtract the whole numbers.

$$\begin{array}{r} 1^1 2^1 2 \\ -89 \\ \hline 33 \end{array}$$

Look at all the steps to the completed problem. Soon you will discover ways to shortcut your work. But make sure you understand before you begin taking those shortcuts.

$$\begin{array}{rrrrr} 123\frac{1}{3} = & 123\frac{8}{24} = & 122 + 1\frac{8}{24} = & 1^12^12\frac{32}{24} \\ -89\frac{5}{8} = & -89\frac{15}{24} = & -89\frac{15}{24} = & -89\frac{15}{24} \\ \hline & & & 33\frac{17}{24} \end{array}$$

There are many different problems that will involve fractions and mixed numbers. If you understand how to change a mixed number to an improper fraction and an improper fraction to a mixed number, you will be able to solve most of the problems involving them.

Multiplication

To multiply fractions, multiply numerators and write the product as the numerator in the answer, then multiply the denominators and write the product as the denominator in the answer. Reduce if necessary. Examples.

$$\frac{2}{3} \cdot \frac{3}{4} = \frac{6}{12} = \frac{1}{2} \qquad \frac{2}{3} \cdot \frac{3}{4} = \frac{2 \times 3}{3 \times 4} = \frac{6}{12} = \frac{1}{2}$$

The example on the next page shows that more than two fractions can be multiplied in one problem and leads to a demonstration of the process of cancellation.

$$\frac{2}{3} \cdot \frac{3}{4} \cdot \frac{4}{5} = \frac{2 \cdot 3 \cdot 4}{3 \cdot 4 \cdot 5}$$

$$\frac{24}{60} = \frac{2 \cdot 2 \cdot 3 \cdot 2}{2 \cdot 2 \cdot 3 \cdot 5} = \frac{2}{5}$$

The problem demonstrates the multiplication of numerators and denominators to get $\frac{24}{60}$ and the factorization of 24 and 60 to reduce to lowest terms. In the factorization of 24 and 60, 3 ones are separated and removed as shown below.

$$\frac{24}{60} = \left(\frac{2}{2}\right)\left(\frac{2}{2}\right)\left(\frac{3}{3}\right)\left(\frac{2}{5}\right) = 1 \times 1 \times 1 \times \frac{2}{5} = \frac{2}{5}$$

Cancellation is a shortcut to the same product. Cancellation would look like this:

$$\frac{2}{\cancel{3}} \cdot \frac{\cancel{3}^1}{\cancel{4}_1} \cdot \frac{\cancel{4}^1}{5}$$

What really happens is that cancelling shortcuts regrouping and changing to one. Regrouping does this to the problem:

$$\frac{2}{3} \cdot \frac{3}{4} \cdot \frac{4}{5} = \frac{3}{3} \cdot \frac{4}{4} \cdot \frac{2}{5}$$

There is a law in mathematics called the commutative law which allows numbers in multiplication problems (addition problems too) to move about or change places without changing the value of the product. So $\frac{3}{3}$ is the equivalent of 1 and so is $\frac{4}{4}$ the equivalent of 1. If there are numbers in the numerator and denominator equivalent to 1, their removal can simplify solution. Watch for multiples when you look for ones too. Example.

$$\frac{2}{9} \cdot \frac{3}{16} \cdot \frac{4}{7} =$$

Note that the 3 and 9 are related: 9 is a multiple of 3; 16 is a multiple of 2 and 4. The fractions can be regrouped like this:

$$\frac{2}{7} \cdot \frac{3}{9} \cdot \frac{4}{16}$$

Now $\frac{3}{9}$ and $\frac{4}{16}$ can both be reduced. Factor and remove the ones.

$$\frac{3}{9} = \frac{1 \cdot \cancel{3}}{3 \cdot \cancel{3}} = \frac{1}{3} \qquad \frac{4}{16} = \frac{1 \cdot \cancel{4}}{4 \cdot \cancel{4}} = \frac{1}{4}$$

Now the problem reads like this:

$$\frac{2}{7} \cdot \frac{1}{3} \cdot \frac{1}{4}$$

And can be regrouped again as below.

$$\frac{1}{7} \cdot \frac{1}{3} \cdot \frac{2}{4}$$

The $\frac{2}{4}$ is factored and reduced

$$\frac{2}{4} = \frac{1 \cdot 2}{2 \cdot 2} = \frac{1}{2}$$

And the problem becomes

$$\frac{1}{7} \cdot \frac{1}{3} \cdot \frac{1}{2} = \frac{1}{42}$$

It is undoubtedly easier to use cancelling to solve this problem.

$$\frac{2}{9} \cdot \frac{3}{16} \cdot \frac{4}{7} = \frac{\cancel{2}^1}{\cancel{9}_3} \cdot \frac{\cancel{3}^1}{\cancel{16}_{\cancel{4}_2}} \cdot \frac{\cancel{4}^1}{7} = \frac{1}{42}$$

However, factoring and removing the ones will make sense to many who have difficulty understanding the process of cancellation.

In multiplying a whole number by a fraction, remember that a whole number is a numerator over 1. For example:

$$\left(7 = \frac{7}{1}\right) \quad 7 \times \frac{1}{3} = \quad \frac{7}{1} \times \frac{1}{3} \quad \frac{7 \times 1}{1 \times 3} = \frac{7}{3} = 2\frac{1}{3}$$

To multiply mixed numbers change them to improper fractions and proceed as before. Examples.

$$1\frac{2}{3} \times \frac{33}{4} = \frac{5}{3} \times \frac{15}{4} = \frac{75}{12} = 6\frac{3}{12} = 6\frac{1}{4}$$

$$2\frac{5}{6} \times 7\frac{7}{15} = \frac{17}{\cancel{6}_3} \times \frac{\cancel{112}^{56}}{15} = \frac{952}{45} = 21\frac{7}{45}$$

$$7\frac{3}{8} \times \frac{3}{4} = \frac{59}{8} \times \frac{3}{4} = \frac{177}{32} = 5\frac{17}{32} \qquad 9 \times 6\frac{2}{3} = \frac{\cancel{9}^3}{1} \times \frac{20}{\cancel{3}_1} = \frac{60}{1} = 60$$

$$113\frac{2}{3} \times 87\frac{7}{9} = \frac{341}{3} \times \frac{790}{9} = \frac{341 \times 790}{3 \times 9} = \frac{269390}{27} = 9977\frac{11}{27}$$

This last example might be easier to solve vertically.

$$113\frac{2}{3}$$
$$\times 87\frac{7}{9} \qquad \text{Explanation}$$

$$\frac{14}{27} \qquad \left(\frac{7}{9} \times \frac{2}{3}\right)$$

$$87\frac{24}{27} \qquad \left(\frac{7}{9} \times 113\right) \quad \frac{7}{9} \times \frac{113}{1} = \frac{791}{9} = 87\frac{8}{9} = 87\frac{24}{27}$$

$$58 \qquad \left(87 \times \frac{2}{3}\right)$$
$$791 \qquad (7 \times 113)$$
$$9040 \qquad (80 \times 113)$$
$$\overline{9976\frac{38}{27}} = 1\frac{11}{27}$$

$$1\frac{11}{27}$$
$$\overline{9977\frac{11}{27}}$$

Choose your own method. If you have an adding machine or calculator available, it doesn't make too much difference which method you choose. The important thing is to understand what you are doing so that you know the process is correct.

Division

Understanding reciprocals will help with division of fractions. The reciprocal of a number is the factor needed to produce a product of 1. For example, the reciprocal of 3 is $\frac{1}{3}$ because $3 \cdot \frac{1}{3} = 1$. The reciprocal of $\frac{3}{5}$ is $\frac{5}{3}$ because $\frac{3}{5} \times \frac{5}{3} = 1$. The reciprocal of a fraction is the fraction inverted (turned upside down).

The rule of dividing fractions is to invert and multiply. Always invert the fraction to the right of the divided by (÷) sign. Examples.

$$\frac{1}{2} \div \frac{1}{3} = \frac{1}{2} \times \frac{3}{1} = \frac{3}{2} = 1\frac{1}{2} \qquad \frac{3}{4} \div \frac{5}{6} = \frac{3}{\cancel{4}_2} \times \frac{\cancel{6}^3}{5} = \frac{9}{10}$$

A question commonly asked by students is "why do you invert?" Below is another way of stating a division by fractions problem. This method of stating the problem makes it easier to demonstrate why the right-hand fraction is inverted and how reciprocals help to solve the problems. Stated as shown, one-half divided by one-third can be confusing. But if the denominator $\frac{1}{3}$ can be changed to 1 the problem will be simplified. That is exactly what the use of the reciprocal permits us to do. Remember, if both the numerator and denominator of a fraction are multiplied or divided by the same number the answer has the same value as the original fraction.

$$\frac{\frac{1}{2}}{\frac{1}{3}} = \frac{\frac{1}{2} \times \frac{3}{1}}{\frac{1}{3} \times \frac{3}{1}} =$$

$$\frac{\frac{1}{2} \times \frac{3}{1}}{\frac{3}{3}} = \frac{\frac{3}{2}}{\frac{3}{3}} = \frac{1\frac{1}{2}}{1} = 1\frac{1}{2}$$

Now check the steps.
Start with

$$\frac{\frac{1}{2}}{\frac{1}{3}} =$$

Multiply the denominator by its reciprocal.

$$\frac{\frac{1}{2}}{\frac{1}{3} \times \frac{3}{1}}$$

Multiply the numerator by the same number.

$$\frac{\frac{1}{2} \times \frac{3}{1}}{\frac{1}{3} \times \frac{3}{1}} =$$

Carry out the operations.

$$\frac{\frac{3}{2}}{\frac{3}{3}} = \frac{1\frac{1}{2}}{1} = 1\frac{1}{2}$$

Note that this fraction has a denominator equivalent to 1 and that the numerator is the same as when you "invert and multiply." So, to invert and multiply is a math shortcut. Understanding why you can take this shortcut may be important to you.

$$\frac{\frac{1}{2} \times \frac{3}{1}}{\frac{1}{3} \times \frac{3}{1}}$$

To divide a fraction by a whole number, write the whole number in fractional form and follow the usual procedure.

$$\frac{3}{4} \div 9 = \qquad \frac{\frac{3}{4}}{9} = \frac{\frac{3}{4}}{\frac{9}{1}} =$$

$$\frac{3}{4} \div \frac{9}{1} = \qquad$$

$$\frac{3}{4} \times \frac{1}{9} = \qquad \frac{\frac{3}{4} \times \frac{1}{9}}{\frac{9}{1} \times \frac{1}{9}} = \frac{\frac{3}{4} \times \frac{1}{9}}{1} =$$

$$\frac{\cancel{3}^1}{4} \times \frac{1}{\cancel{9}_3} = \frac{1}{12} \qquad \frac{3}{4} \times \frac{1}{9} = \frac{3}{36} = \frac{1}{12}$$

To divide a whole number by a fraction:

$9 \div \frac{3}{4} =$ $\qquad$ $\dfrac{9}{\frac{3}{4}} = \dfrac{\frac{9}{1}}{\frac{3}{4}} =$

$\frac{9}{1} \div \frac{3}{4} = \frac{9}{1} \times \frac{4}{3} =$ $\qquad$ $\dfrac{\frac{9}{1} \times \frac{4}{3}}{\frac{3}{4} \times \frac{4}{3}} = \dfrac{\frac{9}{1} \times \frac{4}{3}}{1} = \frac{9}{1} \times \frac{4}{3} = \frac{36}{3} = 12$

$\frac{3\cancel{9}}{1} \times \frac{4}{\cancel{3}_1} = \frac{12}{1} = 12$

To divide a mixed number by a fraction, change the mixed number to an improper fraction and divide.

$1\frac{3}{8} \div \frac{2}{3} =$ $\qquad$ $\dfrac{1\frac{3}{8}}{\frac{2}{3}} = \dfrac{\frac{11}{8}}{\frac{2}{3}} =$

$\frac{11}{8} \div \frac{2}{3} =$ $\qquad$ $\dfrac{\frac{11}{8} \times \frac{3}{2}}{\frac{2}{3} \times \frac{3}{2}} = \dfrac{\frac{11}{8} \times \frac{3}{2}}{1} =$

$\frac{11}{8} \times \frac{3}{2} =$ $\qquad$ $\frac{11}{8} \times \frac{3}{2} = \frac{33}{16} = 2\frac{1}{16}$

$\frac{33}{16} = 2\frac{1}{16}$

To divide a fraction by a mixed number:

$\frac{2}{3} \div 1\frac{3}{8} =$ $\qquad$ $\dfrac{\frac{2}{3}}{1\frac{3}{8}} = \dfrac{\frac{2}{3}}{\frac{11}{8}} =$

$\frac{2}{3} \div \frac{11}{8} =$ $\qquad$ $\dfrac{\frac{2}{3} \times \frac{8}{11}}{\frac{11}{8} \times \frac{8}{11}} = \dfrac{\frac{2}{3} \times \frac{8}{11}}{1} =$

$\frac{2}{3} \times \frac{8}{11} = \frac{16}{33}$ $\qquad$ $\frac{2}{3} \times \frac{8}{11} = \frac{16}{33}$

To divide a mixed number by a mixed number:

$1\frac{3}{7} \div 2\frac{1}{5} =$ $\qquad$ $\dfrac{1\frac{3}{7}}{22\frac{1}{5}} = \dfrac{\frac{10}{7}}{\frac{11}{5}} =$

$\frac{10}{7} \div \frac{11}{5} =$ $\qquad$ $\dfrac{\frac{10}{7} \times \frac{5}{11}}{\frac{11}{5} \times \frac{5}{11}} = \dfrac{\frac{10}{7} \times \frac{5}{11}}{1} =$

$\frac{10}{7} \times \frac{5}{11} = \frac{50}{77}$ $\qquad$ $\frac{10}{7} \times \frac{5}{11} = \frac{50}{77}$

HOW TO OPERATE WITH DENOMINATE NUMBERS

In foodservice/lodging work you will often encounter denominate numbers. <u>Denominate numbers</u> are those which express units of measure like yards, feet and inches, pounds and ounces, cups, tablespoons, and teaspoons. These operations are explained where they are encountered in the word problems in this workbook, so a quick review should suffice.

ADDITION

To add denominate numbers, arrange and label the unit columns, then add. Simplify the sum so no larger unit is contained in smaller units. Add.

```
1 yard    2 feet    3 inches
          1 foot    2 inches
3 yards            11 inches
4 yards   3 feet   16 inches
```

Simplify by changing the 16 inches to feet and inches.

```
       1 foot        12 inches = 1 foot
    12)16            16 inches = 1 foot 4 inches
       12
        4 inches    REMAINDER
```

Combine the 3 feet and 1 foot to make 4 feet

```
4 yards   3 feet   16 inches
          1 foot    4 inches
4 yards   4 feet    4 inches
```

There are no more larger units in the 4 inches but the 4 feet can be changed to 1 yard and 1 foot and the yards combined

```
4 yards   4̸ feet   4 inches
1 yard    1 foot
5 yards   1 foot    4 inches
```

Thus, 5 yards 1 foot 4 inches is the simplified sum in this problem.

SUBTRACTION

To subtract denominate numbers, arrange in like columns and subtract. Simplify the answer if necessary.

```
   3 gallons   2 quarts
 − 1 gallon    1 quart
   2 gallons   1 quart
```

If you need to borrow, change the unit borrowed to the proper number of the units needed.

    3 gallons   2 quarts
 − 1 gallon   3 quarts

Since you can't take 3 quarts from 2 quarts, borrow 1 gallon from the 3 gallons leaving 2 gallons. Change the borrowed gallon to 4 quarts and add to the 2 quarts making 6 quarts, then subtract.

    2   6
   3̸ gal. 2̸ qt.
 − 1 gal. 3 qt.
   1 gal. 3 qt.

The modern way of saying the same thing would be to regroup 3 gal. 2 qt. into 2 gal. 6 qt. and subtract.

From 1 cup take 3 tablespoons. How many tablespoonsful are left in the cup?

   1 cup   =   16 T.
−   3 tablespoons = − 3 T.
                            13 T.

Change the 1 cup to 16 tablespoons and then subtract the three tablespoons. 13 tablespoonsful are left in the cup.

MULTIPLICATION

To multiply denominate numbers, multiply each unit by the multiplier and simplify the products if necessary.

   3 cups   9 tablespoons
   ×   7
 21 cups 6̸3̸ tablespoons
   3 cups 15 tablespoons
 24 cups 15 tablespoons

Since 16 T. = 1 c. then 63 T. ÷ 16 T. will tell how many cups are in 63 tablespoons

       3 cups
16 ⟌ 63
    48
    15 tablespoons   REMAINDER

If desired, the 24 cups could be expressed in the larger units of pints, quarts, or gallons. Refer to the Supplementary Tables and Charts section for equivalences.

DIVISION

To divide denominate numbers, divide each unit column by the divisor. Change any remainders in a unit column to the smaller unit before continuing division. (See examples.)

      4 lbs. 2 oz.
4 ⟌ 16 lbs. 8 oz.
    16   8

In the problem at the right, follow these steps.

Step one. Change the 1 quart remainder to 2 pints.

Step two. Combine the 2 pints with the 1 pint in the problem.

Step three. Divide 2 into 3.

Step four. Change the remainder into $\frac{1}{2}$ pint.

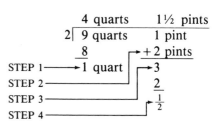

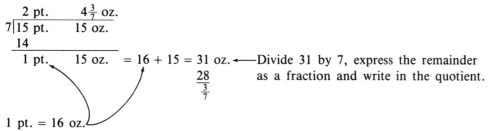

—Divide 31 by 7, express the remainder as a fraction and write in the quotient.

HOW-TO OF RATIO, PROPORTION, AND PERCENT

A rational number is one which may be expressed as the quotient of two whole numbers, except that the divisor cannot be zero. A ratio is a quotient too. A ratio is usually expressed in fractional form, either as $\frac{a}{b}$ or a/b. You can tell whether or not a fraction is being used as a ratio by the way it is used. A ratio is a way of comparing numbers.

Order is important in ratio. If you say the ratio of peas in a pod is 6 to 1 that is certainly not the same as 1 to 6. One pea in six pods would be a poor crop! Be sure the order of your terms correctly states the situation.

Proportion can be a powerful tool in foodservice/lodging mathematics. Proportion is a statement of equality between two ratios. Therefore, proportion can be the key to making solution of percentage simple.

Rational numbers in the form of $\frac{a}{b}$ or a/b are called <u>fractions</u> or <u>ratios</u>. But rational numbers represent percent too. All of these—fractions, ratios, and percents—name quotients. It follows that percent is a special kind of ratio. It is special because 100 is <u>always</u> the basis for comparison. Remember percent means "per hundred" and the symbol % stands for "per hundred." So 10% stands for $\frac{10}{100}$ or any of the equivalent fractions or ratios like $\frac{1}{10}$, $\frac{2}{20}$, $\frac{3}{30}$, $\frac{1}{2} \div 5$, $\frac{100}{1000}$, and so on. In like manner, it can be seen that any fraction or ratio can be expressed as a percent by changing it to an equivalent fraction with 100 as the denominator. Thus the following equation states the proportion equivalent to percent.

$$\frac{a}{b} = \frac{c}{100}$$

The numerator "c" can always be written "c percent" or "c%."
Example:

$$\frac{a}{b} = \frac{c}{100} \quad \text{If} \quad \frac{a}{b} = 1\frac{1}{10} \quad \text{then} \quad \frac{1}{10} = \frac{c}{100}.$$

Solving for c:

$$10c = 1 \cdot 100 \quad c = \frac{100}{10} = 10$$

Therefore c = 10 and the equation becomes $\frac{1}{10} = \frac{10}{100}$. c is then equivalent to 10%.

When we say 10%, or ten percent, what we are communicating is a rate or ratio. 10% is 10 out of 100, or a ratio of 1 to 10.

<u>Ratio</u> is another way to show a division problem or express a quotient. Therefore, every fraction expresses a ratio. For example, ¼ means you have one out of four parts. To express that fraction as a ratio we say it is in the ratio of one to four. We can write it as $\frac{1}{4}$, 1:4, 1/4, 1 ÷ 4, 4⟌1, or one divided by four. So almost any division problem can be read as a ratio. The ratio is the quotient.

If we express a relationship where two fractions are equivalent, or where two division problems have the same answer we are expressing a proportion. A proportion is a number sentence that says one ratio is equal in value to another ratio. For example, we are stating a proportion when we say $\frac{1}{4} = \frac{2}{8}$. Other ways of saying the same thing would include: 1:4 = 2:8 (Read: one is to four as two is to eight) and 1/4 = 2/8.

The four numerals used in a proportion statement are called terms, just as they are in fractions. A problem exists when one of the terms is unknown. In such a case, you can substitute a symbol, such as "x", for the missing term and solve for x. Examples:

$$\frac{1}{4} = \frac{2}{8} \quad \frac{x}{4} = \frac{2}{8} \quad \frac{1}{x} = \frac{2}{8} \quad \frac{1}{4} = \frac{x}{8} \quad \frac{1}{4} = \frac{2}{x}$$

In the last four examples above, we have substituted x for one of the terms. You can see what x stands for by looking at the first example. However, if you didn't have example 1 to look at, you might have to solve a problem to determine the value of x.

In order to solve problems like this, mathematicians have agreed to call two terms in a proportion the "means" and the other two the "extremes." The terms are also numbered. We shall use four letters to help you understand the terms and the numbering system. A proportion statement using four letters follows.

$$\frac{a}{b} = \frac{c}{d}$$

Remember: Read this "a is to b as c is to d." a is the first term. b is the second term. c is the third term. d is the fourth term.

The first and fourth terms are called the "extremes" and the second and third terms are called the "means." When we write the proportion as we did below, it may be easier to understand the use of the words "means" and "extremes."

```
      MEANS
      ⌒
   a:b = c:d
      ⌣
    EXTREMES
```

If you know that "means" is a word that can stand for the words "in the middle" and that "extremes" is another word for "at the utmost point" or "at the ends," it will all make sense.

To solve a proportion, you follow a rule which uses the words means and extremes. The rule is: "The product of the means equals the product of the extremes." In our statement using letters (a:b = c:d) that rule says b × c = a × d. If you are writing proportions in fraction form it is necessary that you know where the means and extremes are. Thus, in $\frac{a}{b} = \frac{c}{d}$, the terms b and c are the means and the terms a and d are the extremes. Another way of stating our rule then is that term 1 multiplied by term 4 equals term 2 multiplied by term 3. The product of term 1 and term 4 equals the product of term 2 and term 3.

Another way of being sure the products are correct is to use the cross multiplication process. This is by far the easiest method of solving proportion and percentage problems. Proportion is an expression of equality between two ratios. Therefore, if we multiply each ratio by 1, expressed in terms of the denominator of the other ratio, we should end up with exactly the same fractions. For example, given the proportion

$$\frac{3}{4} = \frac{6}{8}$$

Each ratio multiplied by 1 in terms of the denominator of the other ratio gives us

$$\frac{3}{4} \times \frac{8}{8} = \frac{4}{4} \times \frac{6}{8}$$

which produces exactly the same fractions.

$$\frac{24}{32} = \frac{24}{32}$$

If the fractions do not come out exactly the same, either a mistake has been made, or the statement is not a proportion: the terms are not equal.

Do you see that when the terms are multiplied as they are above that the equivalency of the proportion is proven? Try the shortcut, cross multiplication, and you will have the same result.

The proportion

$$\frac{3}{4} = \frac{6}{8}$$

cross multiplied gives numbers equal to the numerators in the problem above.

$$\frac{3}{4} \times \frac{6}{8} = \frac{24}{24}$$

When you look at the equation $\frac{a}{b} = \frac{c}{d}$, you can see that cross multiplication is an easy way to remember $a \times d = b \times c$ and eliminates all need to refer to the "means and extremes" as a method of solving proportion and percent problems.

If, then, we know three terms but do not know a fourth term, we should be able to find the missing term using our rule or cross multiplying. For example:

Example one.

If $\quad \frac{x}{4} = \frac{2}{8}$ or x:4 = 2:8

then $\quad x \cdot 8 = 4 \cdot 2$
$\quad\quad\quad 8x = 8$
$\quad\quad\quad x = 8 \div 8$
$\quad\quad\quad x = 1$

Example two.

If $\quad \frac{1}{x} = \frac{2}{8}$ or 1:x = 2:8

then $\quad 1 \cdot 8 = 2 \cdot x \quad$ or $\quad 1 \cdot 8 = 2x$
$\quad\quad\quad 8 = 2x \quad\quad\quad\quad\quad\quad 8 = 2x$
$\quad\quad\quad 8 \div 2 = x \quad\quad\quad\quad\quad \frac{8}{2} = x$
$\quad\quad\quad 4 = x \quad\quad\quad\quad\quad\quad\quad 4 = x$

Example three.

If $\quad \frac{1}{4} = \frac{x}{8}$ or 1:4 = x:8

then $\quad 1 \cdot 8 = 4 \cdot x \quad$ or $\quad 1 \cdot 8 = 4x$
$\quad\quad\quad 8 = 4x \quad\quad\quad\quad\quad\quad 8 = 4x$
$\quad\quad\quad 8 \div 4 = x \quad\quad\quad\quad\quad \frac{8}{4} = x$
$\quad\quad\quad 2 = x \quad\quad\quad\quad\quad\quad\quad 2 = x$

Example four.

If $\quad \frac{1}{4} = \frac{2}{x} \quad$ or $\quad$ 1:4 = 2:x

$\quad\quad 1 \cdot x = 4 \cdot 2 \quad\quad 1x = 4 \cdot 2$
$\quad\quad 1x = 8 \quad\quad\quad\quad\quad 1x = 8$
$\quad\quad x = 8 \div 1 \quad\quad\quad x = \frac{8}{1}$
$\quad\quad x = 8 \quad\quad\quad\quad\quad\quad x = 8$

In the foodservice/lodging industry proportion serves a great need. For example, many times it is the easiest and most efficient method of pricing recipes. For example, if flour costs 14 cents a pound, how much does 14 ounces of flour cost? The proportion is set up as follows, $\frac{14}{16} = \frac{x}{14}$. This says 14 cents is to 16 ounces as x cents is to 14 ounces. Notice, the terms were changed so the units are the same, i.e., one pound was changed to 16 ounces.

Solve the proportion. $16x = 14 \cdot 14$

$$x = \frac{\overset{7}{\cancel{14}} \cdot \overset{7}{\cancel{14}}}{\underset{\underset{4}{8}}{\cancel{16}}} = \frac{49}{4} = 12\frac{1}{4}$$

$x = 12\frac{1}{4}$ cents

Since all the numbers were even, the problem was kept in this form and cancelled before completing the multiplication and division. By this rather simple means we find 14 oz. of flour costs 12¼¢.

Sometimes a simple sounding problem can result in a complicated answer. For example, take the following two questions both of which are common in the foodservice industry. If it costs $2.97 to bake 14 loaves of bread, how much does it cost to bake one loaf? If the bread sells for 35¢ a loaf, what percent of the selling price is profit?

Take one question at a time. Figure out the cost of one loaf from this proportion. Cross multiply. Divide to get the cost of one loaf, $\$.21\frac{3}{14}$.

$$\frac{14}{\$2.97} = \frac{1}{x} \quad 14x = \$2.97 \cdot 1$$
$$x = \frac{\$2.97}{14} = \$.21\frac{3}{14}$$

Knowing that the cost of one loaf of bread is $\$.21\frac{3}{14}$ and the selling price is $.35, makes it possible to find how much profit is made.

$$\begin{array}{r} \$.35 \\ -.21\frac{3}{14} \\ \hline \$.13\frac{11}{14} \text{ PROFIT} \end{array}$$

The question was, "What percent of the selling price is profit?" Remember that percent is per hundred and set up the proportion. Cross multiply. The steps below progress one step at a time in order to make the operations easier to follow. The final answer is $x = 39\frac{19}{49}$. When this number is placed in the proportion, it has a denominator of 100 and can be read $39\frac{19}{49}$ percent.

$$\frac{x}{100} = \frac{13\frac{11}{14}}{35}$$

$$35x = 100 \cdot 13\frac{11}{14}$$

$$35x = 100 \cdot \frac{193}{14}$$

$$35x = \frac{19300}{14}$$

$$x = \frac{19300}{14} \div \frac{35}{1}$$

$$x = \frac{19300}{14} \times \frac{1}{35}$$

$$x = \frac{19300}{490}$$

$$x = 39\frac{19}{49}$$

To prove the answer rewrite the proportion and cross multiply.

$$\frac{x}{100} = \frac{13\frac{11}{14}}{35} \qquad \frac{39\frac{19}{49}}{100} = \frac{13\frac{11}{14}}{35}$$

$$39\frac{19}{49} \cdot 35 = 100 \cdot 13\frac{11}{14}$$

$$1378\frac{4}{7} = 1378\frac{4}{7}$$

Look at one more problem before leaving this area. One cup of allspice weighs 4 oz. and costs $2.28. How much does 1 tablespoon cost? (1c. = 16T.)

Set up the proportion

$$\frac{16}{\$2.28} = \frac{1}{x}$$

Cross multiply

$$16x = \$2.28 \cdot 1$$
$$16x = \$2.28$$

Divide by 16 to find x.

$$x = \frac{\$2.28}{16} = \$.14\frac{1}{4}$$

The cost of one tablespoon of allspice is 14¼ ¢ at the rate of the above price.

Now, what is the relationship between ratio, proportion, and percentage? How can ratio and proportion help you with percentage problems?

When you understand that the word "percent" means per hundred or hundredths, you will see that percent is a fraction with the denominator one hundred. So 12% is the same as $\frac{12}{100}$ or 12/100. If you know one more number in a percentage problem you can solve for the missing number. If we take our example problem (1:4 = 2:8) and change the one-fourth to $\frac{25}{100}$, we can express it as 25%. In making the proportion into a percentage problem be very careful to get the relationships correct. Thus:

$$1:4 = 2:8 \quad \text{or} \quad \frac{1}{4} = \frac{2}{8}$$

$$\frac{1}{4} \times 8 = 2$$

$$\frac{25}{100} \times 8 = 2$$

$$25\% \text{ of } 8 = 2$$

Expressing the example in a proportion

25% of 8 =

$$\frac{25}{100} = \frac{x}{8} \qquad \text{or} \qquad 25:100 = x:8$$

$$25 \times 8 = 100x \qquad\qquad 25 \cdot 8 = 100x$$
$$200 = 100x \qquad\qquad 200 = 100x$$
$$200 \div 100 = x \qquad\qquad \frac{200}{100} = x$$
$$2 = x \qquad\qquad 2 = x$$

If we state our problem as 25% of what number is 2 then:

$$\frac{25}{100} = \frac{2}{x} \quad \text{or} \quad 25:100 = 2:x$$

$$25 \cdot x = 100 \cdot 2 \qquad\qquad 25x = 100 \cdot 2$$
$$25x = 200 \qquad\qquad 25x = 200$$
$$x = 200 \div 25 \qquad\qquad x = \frac{200}{25}$$
$$x = 8 \qquad\qquad x = 8$$

If we want to find what percent one number is of another number, we would proceed by first letting x represent our percent numeral and placing it over 100 (percent means hundredths) and writing the proportion

$$\frac{x}{100} = \frac{2}{8} \quad \text{or} \quad x:100 = 2:8$$

then solving for x.

$$\frac{x}{100} = \frac{2}{8} \quad \text{or} \quad x:100 = 2:8$$

$$x \cdot 8 = 100 \cdot 2 \qquad\qquad 8x = 2 \cdot 100$$
$$8x = 200 \qquad\qquad 8x = 200$$
$$x = 200 \div 8 \qquad\qquad x = \frac{200}{8}$$
$$x = 25 \qquad\qquad x = 25$$

Substitute the 25 for the x in the equation and you have $\frac{25}{100}$ which is equal to 25%. So, in the problem, ☐% of 8 = 2 we have the answer 25% of 8 = 2.

Solving problems by use of ratio and proportion is efficient and easy. Knowledge of these tools and how to use them can be most helpful to you. Try solving the percentage problems in the Exercises chapter of this book using these tools. With a little practice you may become a math wizard.

EXERCISES

ADDITION

Aa1. *Directions:* Add. Make an answer column on the right-hand side of your paper. Show all of your work.

A. 7 +2	B. 3 +4	C. 4 +5	D. 2 +6	E. 2 +3	F. 9 +7	G. 9 +9	H. 2 +7	I. 7 +6
J. 6 +5	K. 5 +4	L. 6 +7	M. 6 +2	N. 4 +3	O. 7 +9	P. 8 +7	Q. 9 +4	R. 4 +7

S. 3 T. 9 U. $1 + 2 + 4 =$ V. $3 + 5 + 8 =$ W. $4 + 6 + 9 =$ X. $2 + 1 + 4 + 3 + 7 =$
   4   7
   2   5
           3

Aa2.

A. $9 + 8 + 1 + 2 + 7 =$ B. $6 + 7 + 7 + 9 =$ C. $2 + 1 + 3 + 9 + 8 =$ D. $4 + 3 + 4 + 5 =$

E. $3 + 2 + 6 + 7 =$ F. $4 + 5 + 6 + 7 =$ G. 7   H. 9   I. 32   J. 123
                                                                               +6   +8   +24   +231

K. 234 +752	L. 1135 +3212	M. 3578 +4211	N. 6942 +3001	O. 34897 +25102	P. 20152 +34847

Q. 33337   R. 112   S. 777   T. 606   U. $101 + 27 + 31 + 300 =$
   +45622   222   101   21
                 341   21   2
                 224   30
                                       220

Directions: Add. Place answers in answer column on the right-hand side of your paper. Show all your work.

Ab.

A. 12 +18	B. 27 +44	C. 39 +13	D. 56 +37	E. 29 +22	F. 69 +19	G. 456 +237	H. 49 +98

I.	234 +777	J.	234 +432	K.	431 +183	L.	540 +929	M.	504 +929	N.	5001 +6009	O.	201 +999

P. 211
    305
    101
    429

Q. 343
    209
    590
    2000

R. 1 + 27 + 239 + 987 =

S. 21 + 302 + 67 + 201 =

T. 37 + 2001 + 909 + 22 =

Ac. Be sure you use your dollar sign and decimal point.

A. $.01
   .02
   .04
   .03

B. $.06
  +.05

C. $.21
   .26
   .37

D. $.13
   .09
   .21
   .56
   .19

E. $.39
   .47
   .59

F. $.89
   .75
   .68
   .94

G. $.67
   .28
   .36

H. $1.16
  +2.31

I. $1.11
   9.95
   8.75
   3.24
   2.34

J. $ 23.10
   324.59
   678.98
     3.24
     2.37

K. $2001.00
   998.39
    98.38
  1002.60

L. $1.12 + $1.24 + $1.05 + $3.67 =

M. $1.01 + $100.00 + $9.09 =

N. $.03 + $5. + $15.31 =

O. $.08 + $6. + $39.98 + $42.83 =

MULTIPLICATION

Directions: Multiply. Place the answers in an answer column on the right-hand side of your paper. Show all your work.

Ma.

A.	3 ×4	B.	6 ×7	C.	8 ×9	D.	7 ×3	E.	9 ×4	F.	23 ×3	G.	321 ×2	H.	434 ×2	I.	31 ×7

J.	711 ×7	K.	703 ×3	L.	501 ×5	M.	6 ×9	N.	800 ×7	O.	901 ×8	P. 23123 × 3 =

Mb.

A.	1234 ×9	B.	34034 ×8	C.	567 ×7	D.	5001 ×6	E.	1432 ×4	F.	506 ×3	G.	689 ×9	H.	537 ×8

I.	5037 ×9	J.	9821 ×8	K.	8943 ×7	L.	9486 ×6	M.	54021 ×5	N.	34089 ×5	O.	12345679 ×9

Mc.

A. 243 × 21 B. 304 × 12 C. 987 × 11 D. 32 × 43 E. 33 × 33 F. 32 × 44 G. 303 × 22

H. 91 × 89 =   I. 602 × 34 =   J. 7007 × 10 =   K. 900 × 13 =   L. 80 × 67 =

Directions: Multiply. Place products in an answer column on the right-hand side of your paper. Show all your work.

Md.

A. 98057 × 11 B. 75089 × 23 C. 6432 × 45 D. 45067 × 77 E. 988 × 86 F. 41067 × 89 G. 70614 × 78

H. 87546 × 56 I. 777 × 46 J. 12345679 × 81 K. 12345679 × 27 L. 8007 × 97

Me.

A. 12345679 × 63 =   B. 3004 × 54 C. 351 × 214 D. 34 × 304 E. 678 × 789 F. 5046 × 387

G. 9043 × 506 H. 12345679 × 126 I. 87012 × 659 J. 9867 × 4358

Mf. Be careful. Use dollar signs and decimal points where needed.

A. $1.23 × 13 B. $9.95 × 24 C. $10.37 × 67 D. 1098 × $.06 E. 876 × $.67 F. $101.03 × 405 G. $6780.00 × 25

H. $38.76 × 3049 I. $45.67 × 808 J. $707.07 × 7568 =

SUBTRACTION

Directions: Subtract. Place the remainders in an answer column on the right-hand side of your paper. Show all your work.

Sa.

A. 9 − 6 B. 8 − 3 C. 7 − 7 D. 8 − 5 E. 4 − 1 F. 10 − 7 G. 13 − 8 H. 9 − 3 I. 13 − 9

J. 19 − 8 K. 22 − 11 L. 23 − 12 M. 34 − 21 N. 98 − 77 O. 76 − 36 P. 100 − 90 Q. 123 − 91

R. 305   S. 123456789   T. 9876543210
   −201   −92143645   −3304530110

Sb.

A. 987   B. 23   C. 67   D. 89   E. 203   F. 113   G. 301   H. 1001
  −354   −9   −58   −79   −99   −98   −202   −903

I. 2001   J. 697   K. 6097   L. 12345   M. 3000   N. 6000   O. 5000
  −678   −548   −548   −9876   −1987   −29   −500

P. 90 − 27 =   Q. 7000 − 700 =   R. 464 − 422 =   S. 200 − 3 =   T. 4789 − 999 =

Sc. Be sure to use the dollar sign and the decimal point where needed.

A. $.09   B. $.19   C. $.37   D. $.51   E. $1.23   F. $11.37   G. $101.00
  −.07   −.08   −.18   −.49   −.67   −10.27   −89.11

H. $1001.99   I. $654.44   J. $.45   K. $400.00   L. $3456.78   M. $1.34
   −28.87   −500.00   −.37   −111.11   −567.89   −.90

N. $101 − $.99 =   O. $1515.15 − $909.09 =

DIVISION

Directions: Divide. Place the quotients in an answer column on the right-hand side of your paper.

Da.

A. 9)18   B. 3)15   C. 6)24   D. 7)567   E. 9)8109   F. 7)2814   G. 6)1884

H. 5)3565   I. 7)756   J. 4)4816   K. 5)570   L. 2)12012   M. 8)5688   N. 9)11106

O. 9)488889   P. 9)7101   Q. 3)1962   R. $\frac{3648}{8} =$ S. 2247 ÷ 7 =   T. 3321 ÷ 9 =

U. 2064 ÷ 8 =   V. $\frac{5187}{7} =$ W. 7)865578   X. 6)61236540

Db. Write remainders as shown (6 r2)

A. 8)9   B. 8)25   C. 9)289   D. 7)722   E. 7)5658   F. 3)2448   G. 5)5678

H. 6)4739   I. 4)147949   J. 8)480987

Dc.

A. 11)8679   B. 21)13734   C. 13)1274   D. 42)13482   E. 56)2632   F. 87)43587

G. 30)28530   H. 77)52899   I. 50)200150   J. 13)4004

Directions: Divide. Place quotients in an answer column on your paper. Show remainders as shown (6 r2). Show all your work.

Dd.

A. 13)27 B. 39)80 C. 45)100 D. 98)903 E. 71)800 F. 85)1250 G. 12)4000

H. 30)333 I. $.13)$39.40 J. 444 ÷ 21 =

De.

A. 321)1926 B. 246)738 C. 987)18753 D. 645)17415 E. 200)24200 F. 307)31621

G. 141,192 ÷ 666 = H. 152,005 ÷ 505 =

Df.

A. 447)46041 B. 587)123270 C. 474)52615 D. 758)6070 E. 321)10600

F. 663)227411

Dg.

A. 12)36039 B. 33)13398 C. 167)10104001 D. 9001)31107456

Dh. Be sure to use dollar signs and decimal points.

A. 3)$.99 B. 4)$1.64 C. 9)$.81 D. $.04)$.16 E. $3.29)$141.47 F. 67)$439.12

G. 981)$88.29 H. $.03)$.37

FRACTIONS

Directions: Place the answers in a column on the right-hand side of your paper.

Fa. Supply the missing numerators and denominators.

A. $\frac{1}{2} = \frac{}{4}$ B. $\frac{3}{4} = \frac{}{24}$ C. $\frac{1}{2} = \frac{3}{}$ D. $\frac{2}{5} = \frac{}{10}$ E. $\frac{1}{7} = \frac{}{49}$ F. $\frac{1}{3} = \frac{}{9}$ G. $\frac{2}{7} = \frac{14}{}$

H. $\frac{4}{36} = \frac{}{54}$ I. $\frac{2}{11} = \frac{}{66}$ J. $\frac{3}{13} = \frac{9}{}$ K. $\frac{2}{3} = \frac{}{6}$ L. $\frac{5}{25} = \frac{}{35}$ M. $\frac{2}{9} = \frac{}{81}$

N. $\frac{3}{15} = \frac{}{45}$ O. $\frac{1}{13} = \frac{}{117}$

Fb. Reduce to lowest terms.

A. $\frac{2}{4} = $ — B. $\frac{6}{8} = $ — C. $\frac{3}{9} = $ — D. $\frac{6}{12} = $ — E. $\frac{3}{33} = $ — F. $\frac{6}{54} = $ — G. $\frac{21}{24} = $ —

H. $\frac{14}{16} = $ — I. $\frac{28}{49} = $ — J. $\frac{16}{48} = $ — K. $\frac{6}{16} = $ — L. $\frac{6}{48} = $ — M. $\frac{8}{80} = $ —

N. $\frac{7}{77} = $ —

Directions: Place answers in a column on the right-hand side of your paper.

Fc1. Change the following to mixed numbers.

A. $\frac{9}{8} =$ B. $\frac{10}{8} = 1\frac{2}{8} = 1\frac{1}{4}$ C. $\frac{4}{3} =$ D. $\frac{7}{4} =$ E. $\frac{11}{9} =$ F. $\frac{7}{3} = 2\text{—}$ G. $\frac{5}{2} =$

H. $\frac{5}{4} =$ I. $\frac{13}{10} =$ J. $\frac{9}{7} =$ K. $\frac{5}{3} =$ L. $\frac{6}{5} =$ M. $\frac{15}{13} =$ N. $\frac{23}{13} =$ O. $\frac{40}{13} =$

P. $\frac{6}{4} =$ Q. $\frac{8}{6} =$ R. $\frac{35}{17} =$ S. $\frac{5}{2} =$ T. $\frac{202}{117} =$

Fc2. Change the mixed numbers below to improper fractions.

A. $1\frac{1}{3} = \text{—}$ B. $1\frac{1}{4} = \text{—}$ C. $1\frac{3}{4} = \text{—}$ D. $1\frac{7}{8} = \text{—}$ E. $1\frac{13}{16} = \text{—}$ F. $1\frac{5}{8} = \text{—}$

G. $1\frac{13}{32} = \text{—}$ H. $1\frac{3}{13} = \text{—}$ I. $1\frac{9}{12} = \text{—}$ J. $1\frac{2}{5} = \text{—}$ K. $1\frac{2}{3} = \text{—}$ L. $1\frac{3}{7} = \text{—}$

M. $2\frac{2}{3} = \text{—}$ N. $3\frac{1}{7} = \text{—}$ O. $3\frac{1}{13} = \text{—}$

Directions: Add. Place the answers in an answer column on the right-hand side of your paper.

Fd.

A. $\frac{1}{3} + \frac{1}{3}$ B. $\frac{2}{5} + \frac{1}{5}$ C. $\frac{3}{7} + \frac{1}{7}$ D. $\frac{1}{5} + \frac{1}{5} + \frac{1}{5}$ E. $\frac{2}{9} + \frac{3}{9}$ F. $\frac{1}{9} + \frac{2}{9} + \frac{4}{9}$ G. $\frac{1}{5} + \frac{1}{5} + \frac{1}{5} + \frac{1}{5}$

H. $\frac{3}{11} + \frac{4}{11}$ I. $\frac{7}{13} + \frac{2}{13}$ J. $\frac{4}{15} + \frac{9}{15}$ K. $\frac{1}{15} + \frac{2}{15} + \frac{6}{15} + \frac{3}{15}$ L. $\frac{9}{23} + \frac{7}{23}$ M. $\frac{2}{17} + \frac{3}{17} + \frac{4}{17} + \frac{5}{17}$ N. $\frac{2}{7} + \frac{2}{7} + \frac{2}{7}$

Fd2. Express all answers in lowest terms.

A. $\frac{1}{3} + \frac{1}{3} =$ B. $\frac{1}{4} + \frac{1}{4} + \frac{1}{4} =$ C. $\frac{1}{9} + \frac{2}{9} + \frac{3}{9} =$ D. $\frac{1}{7} + \frac{2}{7} + \frac{3}{7} =$ E. $\frac{4}{9} + \frac{3}{9} =$

F. $\frac{7}{11} + \frac{3}{11} =$ G. $\frac{2}{17} + \frac{3}{17} + \frac{5}{17} + \frac{1}{17} =$ H. $\frac{2}{7} + \frac{2}{7} + \frac{2}{7} =$ I. $\frac{2}{8} + \frac{1}{8} + \frac{3}{8} =$

J. $\frac{7}{39} + \frac{3}{39} + \frac{2}{39} + \frac{1}{39} =$

Directions: Add. Place answers in an answer column on the right-hand side of your paper. Express all sums in simplest terms.

Fd3.

A. $\dfrac{1}{2}$
   $+\dfrac{1}{3}$

B. $\dfrac{2}{3}$
   $+\dfrac{1}{4}$

C. $\dfrac{1}{2}$
   $\dfrac{1}{3}$
   $+\dfrac{1}{4}$

D. $\dfrac{3}{4}$
   $+\dfrac{3}{8}$

E. $\dfrac{2}{3}$
   $+\dfrac{5}{6}$

F. $\dfrac{4}{5}$
   $+\dfrac{1}{10}$

G. $\dfrac{1}{7}$
   $+\dfrac{1}{8}$

H. $\dfrac{3}{4}$
   $\dfrac{3}{8}$
   $+\dfrac{3}{16}$

I. $\dfrac{2}{3}$
   $\dfrac{5}{6}$
   $+\dfrac{4}{9}$

J. $\dfrac{1}{2}+\dfrac{2}{3}+\dfrac{3}{4}+\dfrac{4}{5}=$

Fd4. Express all sums in simplest form.

A. $\dfrac{1}{6}$
   $+\dfrac{1}{6}$

B. $2\dfrac{1}{4}$
   $+1\dfrac{1}{4}$

C. $1\dfrac{2}{3}$
   $+2\dfrac{3}{4}$

D. $1\dfrac{1}{2}$
   $2\dfrac{1}{3}$
   $+3\dfrac{1}{4}$

E. $4\dfrac{3}{4}$
   $5\dfrac{1}{6}$
   $+6\dfrac{3}{5}$

F. $7\dfrac{2}{9}$
   $+8\dfrac{7}{8}$

G. $9\dfrac{1}{3}$
   $10\dfrac{2}{5}$
   $+11\dfrac{4}{7}$

H. $\dfrac{3}{13}$
   $+\dfrac{2}{3}$

I. $2\dfrac{5}{7}$
   $30\dfrac{7}{8}$
   $+404\dfrac{8}{9}$

J. $\dfrac{1}{2}+\dfrac{3}{8}+\dfrac{7}{9}+\dfrac{5}{13}=$

Fe1.
Directions: Multiply. Copy the problems on your paper. Show all your work. Make an answer column on the right-hand side of your paper.

A. $\dfrac{1}{2}\times\dfrac{1}{2}=$
B. $\dfrac{1}{3}\times\dfrac{1}{3}=$
C. $\dfrac{1}{4}\times\dfrac{1}{4}=$
D. $\dfrac{1}{5}\times\dfrac{1}{5}=$
E. $\dfrac{1}{6}\times\dfrac{1}{6}=$
F. $\dfrac{1}{7}\times\dfrac{1}{7}=$

G. $\dfrac{1}{8}\times\dfrac{1}{8}=$
H. $\dfrac{1}{9}\times\dfrac{1}{9}=$
I. $\dfrac{1}{10}\times\dfrac{1}{10}=$
J. $\dfrac{1}{11}\times\dfrac{1}{11}=$

Pay close attention to the process in solving the problems above. You will use the same process in all multiplication of fractions problems. Can you find a pattern in the problems above?

Fe2. *Directions:* (Same as in Fe1.)

A. $\dfrac{1}{2}\times\dfrac{1}{3}=$
B. $\dfrac{1}{3}\times\dfrac{1}{4}=$
C. $\dfrac{1}{4}\times\dfrac{1}{5}=$
D. $\dfrac{1}{5}\times\dfrac{1}{6}=$
E. $\dfrac{1}{6}\times\dfrac{1}{7}=$
F. $\dfrac{1}{7}\times\dfrac{1}{8}=$

G. $\dfrac{1}{8}\times\dfrac{1}{9}=$
H. $\dfrac{1}{9}\times\dfrac{1}{10}=$
I. $\dfrac{1}{10}\times\dfrac{1}{11}=$
J. $\dfrac{1}{11}\times\dfrac{1}{12}=$

Fe3. *Directions:* Multiply. Copy the problems on your paper. Show all your work. Make an answer column on the right-hand side of your paper. Express products in lowest terms.

A. $\frac{2}{3} \times \frac{1}{4} =$ B. $\frac{3}{4} \times \frac{1}{5} =$ C. $\frac{2}{5} \times \frac{1}{6} =$ D. $\frac{3}{5} \times \frac{1}{6} =$ E. $\frac{4}{5} \times \frac{1}{6} =$ F. $\frac{1}{2} \times \frac{1}{7} =$

G. $\frac{1}{3} \times \frac{1}{7} =$ H. $\frac{2}{6} \times \frac{1}{7} =$ I. $\frac{3}{6} \times \frac{1}{7} =$ J. $\frac{5}{6} \times \frac{1}{7} =$ K. $\frac{2}{7} \times \frac{2}{8} =$ L. $\frac{3}{7} \times \frac{4}{8} =$

M. $\frac{4}{7} \times \frac{5}{9} =$ N. $\frac{5}{8} \times \frac{7}{9} =$ O. $\frac{7}{8} \times \frac{8}{9} =$

Fe4. *Directions:* (Same as in Fe3.)

A. $\frac{1}{4} \times 1 =$ B. $\frac{1}{2} \times 2 =$ C. $\frac{2}{3} \times 3 =$ D. $\frac{3}{5} \times 4 =$ E. $\frac{5}{6} \times 6 =$ F. $7 \times \frac{4}{7} =$

G. $\frac{5}{8} \times 10 =$ H. $\frac{5}{8}$ of $14 =$ I. $17 \times \frac{11}{12} =$ J. $\frac{2}{3} \times \frac{2}{3} =$

Fe5. Express products in simplest form.

A. $\frac{4}{3} \times \frac{6}{5} =$ B. $\frac{3}{2} \times \frac{3}{2} =$ C. $\frac{1}{2} \times \frac{2}{3} =$ D. $1\frac{1}{2} \times 1\frac{1}{2} =$ E. $2\frac{2}{3} \times 2\frac{3}{4} =$

F. $9\frac{4}{5} \times 3\frac{4}{7} =$ G. $5\frac{2}{5} \times 6\frac{3}{5} =$ H. $19\frac{7}{12} \times 17\frac{4}{17}$ I. $3\frac{4}{9} \times 4\frac{5}{7} =$ J. $2\frac{1}{2} \times 6\frac{2}{3} =$

Ff. *Directions:* All tests with the key "Ff" are subtraction of fractions problems. Copy the problems for each test on a separate paper. Show all your work. Express your remainders or differences in lowest terms, or in simplest form.

Ff1. Subtract.

A. $\frac{2}{3} - \frac{1}{3} =$ B. $\frac{3}{4} - \frac{1}{4} =$ C. $\frac{8}{8} - \frac{5}{8} =$ D. $\frac{7}{8} - \frac{1}{8} =$ E. $\frac{7}{9} - \frac{2}{9} =$ F. $\frac{15}{16} - \frac{3}{16} =$

G. $\frac{5}{6} - \frac{2}{6} =$ H. $\frac{7}{9} - \frac{4}{9} =$ I. $\frac{11}{12} - \frac{1}{12} =$ J. $\frac{7}{13} - \frac{5}{13} =$

Ff2. Subtract.

A. $\frac{2}{3} - \frac{1}{2} =$ B. $\frac{2}{3} - \frac{1}{4} =$ C. $\frac{2}{3} - \frac{1}{5} =$ D. $\frac{5}{9} - \frac{1}{2} =$ E. $\frac{5}{6} - \frac{2}{3} =$ F. $\frac{15}{16} - \frac{5}{8} =$

G. $\frac{5}{8} - \frac{1}{2} =$ H. $\frac{3}{4} - \frac{1}{2} =$ I. $\frac{4}{5} - \frac{8}{15} =$ J. $\frac{7}{8} - \frac{2}{3} =$

Ff3. Subtract.

A. $1 - \frac{1}{2} =$ B. $1 - \frac{1}{4} =$ C. $1 - \frac{1}{5} =$ D. $1 - \frac{1}{6} =$ E. $1\frac{1}{4} - \frac{3}{4} =$ F. $1\frac{1}{2} - \frac{3}{4} =$

G. $1\frac{3}{4} - \frac{7}{8} =$ H. $2 - \frac{1}{2} =$ I. $1\frac{9}{10} - \frac{4}{5} =$ J. $1\frac{1}{2} - \frac{14}{15} =$

Ff4. Subtract.

A. $1\frac{1}{2}$ B. $1\frac{3}{4}$ C. $1\frac{5}{8}$ D. $1\frac{9}{10}$ E. $1\frac{1}{2}$ F. $1\frac{1}{2}$ G. $1\frac{1}{3}$ H. $2\frac{3}{4}$
$-\frac{1}{2}$ $-\frac{1}{2}$ $-\frac{3}{8}$ $-\frac{4}{5}$ $-\frac{4}{5}$ $-\frac{3}{4}$ $-\frac{3}{5}$ $-\frac{7}{8}$

I. $8\frac{2}{5}$ J. 5
$-2\frac{3}{5}$ $-\frac{3}{4}$

Ff5. Subtract.

A. $34\frac{5}{8}$ B. $33\frac{5}{8}$ C. $354\frac{3}{4}$ D. $9\frac{7}{8}$ E. $24\frac{1}{2}$ F. $305\frac{1}{4}$
$-15\frac{1}{8}$ $-9\frac{1}{4}$ $-295\frac{7}{8}$ $-3\frac{15}{16}$ $-12\frac{3}{4}$ $-9\frac{3}{8}$

G. $13\frac{3}{4} - 9\frac{1}{8} =$ H. $20\frac{1}{4} - 9\frac{3}{16} =$ I. $15\frac{5}{6} - 2\frac{1}{3} =$ J. $10\frac{4}{9} - 5\frac{4}{7} =$

Ff6. Subtract.

A. $9\frac{13}{14}$ B. $13\frac{23}{24}$ C. $10\frac{3}{39}$ D. $7\frac{50}{64}$ E. $13\frac{3}{13}$ F. $2\frac{4}{11}$ G. $13\frac{7}{9}$
$-6\frac{11}{12}$ $-9\frac{13}{48}$ $-6\frac{7}{13}$ $-3\frac{3}{8}$ $-9\frac{27}{39}$ $-1\frac{9}{22}$ $-6\frac{5}{6}$

H. $5\frac{5}{7}$ I. $4\frac{9}{16}$ J. $101\frac{1}{16}$
$-3\frac{4}{5}$ $-3\frac{5}{12}$ $-99\frac{1}{17}$

Fg *Directions:* All tests with the key "Fg" are division of fractions problems. Copy the problems for each test on a separate paper. Show all your work. Express your quotients in lowest terms, or in simplest form.

Fg1. Divide.

A. $\frac{1}{2} \div \frac{1}{2} =$ B. $\frac{1}{4} \div \frac{1}{4} =$ C. $\frac{1}{8} \div \frac{1}{8} =$ D. $\frac{1}{3} \div \frac{1}{3} =$ E. $\frac{1}{5} \div \frac{1}{5} =$ F. $\frac{1}{16} \div \frac{1}{16} =$

G. $\frac{1}{9} \div \frac{1}{9} =$ H. $\frac{1}{10} \div \frac{1}{10} =$ I. $\frac{1}{102} \div \frac{1}{102} =$ J. $\frac{3}{4} \div \frac{3}{4} =$ K. $\frac{3}{8} \div \frac{3}{8} =$ L. $\frac{4}{9} \div \frac{4}{9} =$

Fg2. Divide.

A. $\frac{1}{2} \div \frac{2}{1} =$ B. $\frac{1}{2} \div 2 =$ C. $\frac{1}{4} \div \frac{1}{4} =$ D. $\frac{1}{4} \div \frac{1}{2} =$ E. $\frac{1}{4} \div \frac{3}{4} =$ F. $\frac{1}{4} \div 1 =$

G. $\frac{2}{5} \div \frac{2}{5} =$ H. $\frac{2}{5} \div \frac{3}{5} =$ I. $\frac{2}{5} \div \frac{4}{5} =$ J. $\frac{2}{5} \div \frac{5}{5} =$

Fg3. Divide.

A. $\frac{2}{3} \div \frac{3}{4} =$ B. $\frac{3}{4} \div \frac{2}{3} =$ C. $\frac{4}{5} \div \frac{2}{3} =$ D. $\frac{4}{5} \div \frac{3}{4} =$ E. $\frac{2}{5} \div \frac{4}{5} =$ F. $\frac{4}{5} \div \frac{2}{5} =$

Continued on next page.

G. $\dfrac{2}{5} \div \dfrac{2}{9} =$ H. $\dfrac{2}{5} \div \dfrac{5}{9} =$ I. $\dfrac{2}{5} \div \dfrac{4}{9} =$ J. $\dfrac{2}{5} \div \dfrac{7}{9} =$

Fg4. Divide.

A. $1 \div \dfrac{1}{4} =$ B. $1 \div \dfrac{3}{4} =$ C. $\dfrac{3}{2} \div \dfrac{3}{4} =$ D. $1\dfrac{1}{2} \div \dfrac{2}{3} =$ E. $2\dfrac{2}{3} \div \dfrac{2}{3} =$ F. $2\dfrac{1}{2} \div \dfrac{2}{3} =$

G. $\dfrac{3}{4} \div \dfrac{1}{2} =$ H. $3\dfrac{3}{4} \div \dfrac{7}{9} =$ I. $1\dfrac{3}{4} \div \dfrac{2}{7} =$ J. $9\dfrac{1}{9} \div \dfrac{3}{4} =$

Fg5. Divide.

A. $\dfrac{3}{4} \div 2 =$ B. $\dfrac{2}{9} \div 3 =$ C. $3\dfrac{2}{9} \div 2 =$ D. $1\dfrac{1}{4} \div \dfrac{3}{4} =$ E. $1\dfrac{1}{4} \div 2\dfrac{1}{2} =$

F. $2\dfrac{1}{2} \div 1\dfrac{1}{4} =$ G. $3\dfrac{2}{9} \div 2\dfrac{1}{4} =$ H. $\dfrac{7}{9} \div 3\dfrac{3}{4} =$ I. $\dfrac{3}{4} \div 9\dfrac{1}{9} =$ J. $2 \div 3\dfrac{3}{4} =$

Fg6. Divide.

A. $1\dfrac{1}{4} \div 2\dfrac{1}{2} =$ B. $\dfrac{3}{4} \div 1\dfrac{1}{4} =$ C. $2\dfrac{1}{2} \div 3\dfrac{3}{4} =$ D. $2\dfrac{4}{9} \div 5\dfrac{1}{3} =$ E. $6\dfrac{1}{4} \div 7\dfrac{2}{3} =$

F. $7\dfrac{2}{3} \div 6\dfrac{1}{4} =$ G. $26\dfrac{1}{13} \div 3\dfrac{4}{39} =$ H. $13\dfrac{1}{4} \div 17\dfrac{1}{3} =$ I. $9\dfrac{1}{5} \div 9\dfrac{2}{5} =$ J. $9\dfrac{2}{5} \div 9\dfrac{1}{5} =$

DECIMALS

Xa1. *Directions:* Write the following in numerals. Express fractions in decimal form.

A. One hundred one._____   B. Three hundred eleven._____   C. One thousand one._____

D. Two million twenty-one._____   E. One thousand and one-tenth._____

F. One hundred and one-hundredth._____

G. One thousand and one hundred thirty-seven thousandths._____

H. Three trillion three million three._____   I. Five million and thirty-three ten-thousandths._____

J. One and twenty-three hundredths._____

Xb1. Add.

A. $.5 + .4 =$ B. $.1 + .2 + .3 =$ C. $.1 + .2 + .3 + .4 =$ D. $.5 + .6 =$

```
E.  .9     F.  .3     G.  .9     H. 1.1     I. 5.6     J. 3.9
    .8         .2         .6        2.3        6.5        2.1
    .7         .5         .5        3.4        3.5        6.8
    .6         .6                              2.7        7.4
                                                          8.6
```

Xb2. Add.

A. 3.03
  +4.09

B. 30.3 + 4.09 =

C. 3.03
  +40.9

D. 1.1 + .9 =

E. .7 + 7.7 + .77 =

F. .09 + 9.9 + 9.09 =

G. .001 + 1000.1 + .01 =

H. 1.01
  .88
  3.09
  24.11

I. 10.1
  8.8
  324.9
  101.9

J. 129.0
  20.9
  302.2
  1001.6

Xb3. Add.

A. 129 + .19 =

B. 129 + 1.9 =

C. 12.9 + .19 =

D. .129 + .19 =

E. 109 + 3.02 =

F. 1.09 + 3.02 =

G. 123.9 + .04 + 23.1 + 15.99 =

H. 1.239 + .06 + 23.1 + 32 =

I. 100 + 1.01 + 9.9 =

J. 2001 + 99.09 + .011 =

Xb4. Add.

A. Add fifteen cents to $3.

B. Add eight cents to $1.98.

C. Add $1.97 to 33 cents.

D. $.45 + $.59 + $.63 + $.32 =

E. $101.33 + $303.97 + $1 =

F. $3 + $9 + $.46 =

G. $15.15 + $.69 + $1.03 =

H. $101.01 + $.99 =

I. $14.56 + $15.67 + $101. =

J. $1001 + $9.01 =

Xc1. Multiply.

A. 8
  ×9

B. .8
  ×9

C. .7
  ×8.

D. 6.
  ×.5

E. .05
  ×6

F. .05
  ×.6

G. .15
  ×1.1

H. .15
  ×.11

I. .22
  ×.23

J. 2.02
  ×.11

Xc2.

A. 769
  ×.35

B. 96.7
  ×3.8

C. 9.9
  ×.77

D. 80.8
  ×6.6

E. 434
  ×.22

F. 4.034
  ×.9

G. 2.046
  ×2.02

H. 6006.
  ×.34

I. 6006.
  ×.304

J. .909
  ×.707

Xc3.

A. .003 × 3 B. .67 × .003 C. 607 × .003 D. .567 × 1.03 E. .567 × 33 F. 4.07 × 6.09 G. $36.05 × 15

H. $101.95 × 3.09 I. $.45 × .15 J. $10.15 × .05

Xc4. Multiply. Round off products to the nearest hundredth (or the nearest cent).

A. $.46 × .76 B. $1.01 × 7.6 C. $3.33 × .28 D. $3.09 × 1.1 E. $100.39 × .08 F. $133.67 × .09 G. $10.07 × 8.8

H. $15.15 × 37.8 I. $14.03 × 9.9 J. 100.09 × .0625

Xd1. Subtract.

A. 3.3 − 1.1 B. 6.7 − 3.4 C. 8.93 − 2.01 D. 6.03 − 5.23 E. 6.03 − 2.99 F. 78.9 − 23.1 G. 23.9 − 12.7

H. 23.1 − 7.9 I. 20.0 − 9.8 J. 3.07 − 1.08

Xd2. Subtract.

A. $15.00 − $1.08 =

B. $3 − $.08 =

C. $100 − $9.99 =

D. 789 − 8.88 =

E. $10 − $.35 =

F. $1.175 − $.125

G. $3 − $.11 =

H. 1000 − 1.01 =

I. $15.05 − $.15 =

J. 1234 − 456.7 =

Xd3. Subtract.

A. From $5 take 5 cents.

B. From $10 take $1.03.

C. Take eight cents from $1.

D. Take $5.23 from $20.

E. How much is left if you take away fifteen cents from one dollar?

F. What is the remainder of $5 minus $1.98?

G. From $100 take $8.99.

H. Take $.23 away from $5.

I. From $9 take $8.88.

J. Take $6.30 from $10.

Xe1. Divide.

A. .3)‾.6 B. .3)‾6. C. .05)‾10. D. .15)‾.45 E. .15)‾4.5 F. .13)‾39. G. 13)‾.78

H. .5)‾105. I. .05)‾.05 J. .05)‾.055

Xe2. Divide.

A. 1.5⟌45.   B. 1.7⟌170   C. 3⟌.9   D. 22.5⟌45   E. .45⟌9   F. 9⟌.45   G. 1.11⟌1332

H. 9.8⟌68.6   I. .7⟌.49   J. 22.5⟌.225

Xe3. Divide.

A. .45⟌.09   B. 4.5⟌31.5   C. 10⟌5   D. 14.3⟌104.39   E. 15.8⟌$188.34   F. .006⟌.48

G. 4.8⟌9.696   H. .3⟌9   I. 4⟌.0012   J. 14⟌.028

Xe4. Divide. Round off quotients to the nearest cent or hundredth if your answer involves the thousandths column.

A. $.60 ÷ 3 =   B. $1 ÷ 6 =   C. 898 ÷ 3 =   D. $15 ÷ 4 =   E. $15 ÷ $.03 =

F. $90 ÷ $45 =   G. $5 ÷ 3 =   H. $4.44 ÷ 3 =   I. $100 ÷ 15 =   J. $39 ÷ $.13 =

PERCENTAGE

Pa1. Percentage. Change the fractions to equivalent percent numerals.

A. $\frac{1}{2} =$ %   B. $\frac{1}{3} =$ %   C. $\frac{1}{4} =$ %   D. $\frac{1}{6} =$ %   E. $\frac{2}{3} =$ %

F. $\frac{1}{8} =$ %   G. $\frac{3}{4} =$ %   H. $\frac{3}{8} =$ %   I. $\frac{1}{12} =$ %   J. $\frac{3}{6} =$ %

Pa2. Percentage. Change the following percentage figures to fractional equivalents.

A. $12\frac{1}{2}\% =$ B. $37\frac{1}{2}\% =$ C. $33\frac{1}{3}\% =$ D. $16\frac{2}{3}\% =$ E. $62\frac{1}{2}\% =$

F. $50\% =$ G. $25\% =$ H. $6\frac{1}{4}\% =$ I. $8\frac{1}{3}\% =$ J. $87\frac{1}{2}\% =$

Pa3. Percentage. Change the following decimal fractions to percentage equivalents.

A. .125 =   B. .25 =   C. .75 =   D. .385 =   E. .33 =   F. .66 =   G. .33$\bar{3}$ =

H. .66$\bar{6}$ =   I. .0625 =   J. .083$\bar{3}$ =

Pa4. Percentage. Change the following mixed numbers to percentage equivalents.

A. $1\frac{1}{4} =$ %   B. $2\frac{1}{8} =$ %   C. $1\frac{1}{16} =$ %   D. $1\frac{1}{2} =$ %   E. $1\frac{3}{4} =$ %

F. $1\frac{3}{8} =$ %   G. $1\frac{1}{6} =$ %   H. $1\frac{1}{3} =$ %   I. $6\frac{2}{3} =$ %   J. $1\frac{7}{8} =$ %

Pa5. Percentage. Change the following mixed decimals to percentage equivalents.

A. 1.00 = %   B. 1.25 = %   C. 2.5 = %   D. 3.33$\bar{3}$ = %   E. 1.0625 = %

F. 2.125 = %   G. 10.25 = %   H. 3.35 = %   I. $1.66\frac{2}{3} =$ %   J. 1.875 = %

Pb1. Percentage.

A. 10% of 100 =   B. 10% of 10 =   C. 5% of 20 =   D. 5% of 100 =

E. 3% of 100 =   F. 3% of 300 =   G. 100% of 10 =   H. 50% of 10 =

I. 10% of 50 =   J. 15% of 300 =

Pb2. Percentage. Be careful with decimal points.

A. 5% of 10 =   B. 15% of 30 =   C. 9% of 9 =   D. 23% of 50 =   E. 75% of 50 =

F. 25% of 25 =   G. 4% of 100 =   H. 6% of 25 =   I. 3% of 75 =   J. 25% of 96 =

Pb3. Percentage.

A. $12\frac{1}{2}$% of 88 =   B. 25% of 88 =   C. $6\frac{1}{4}$% of 88 =   D. 12.5% of 176 =

E. 6.25% of 176 =   F. $33\frac{1}{3}$% of 99 =   G. 5% of 5.5 =   H. 64% of 25 =

Pb4. Percentage. Find the amount of tax on the following prices using the rates quoted.

A. $5 at 5%   B. $50 at $12\frac{1}{2}$%   C. $100 at $6\frac{1}{4}$%   D. $99 at 3%   E. $33 at $33\frac{1}{3}$%

F. $3 at 5%   G. $32.50 at 8%   H. $50 at 6%   I. $.60 at 10%   J. $1.60 at 5%

Pc1. Percentage.

A. 10% of what number is 10?

B. 5% of what number is 5?

C. 7% of what number is 7?

D. 5% of what number is 3?

E. 15% of what number is 6?

F. 25% of what number is 17?

G. 50% of what number is 15?

H. 75% of what number is 60?

I. 30% of what number is 93?

J. 37% of what number is 111?

Pc2. Percentage. Solve for the missing numbers.

A. 3% of ☐ = 9

B. 5% of n = 20

C. 10% of x = 95

D. 8% · ☐ = 70

E. 15% x = 45

F. 18% of n = 558

G. 95% of ☐ = 19

H. 75% x = 30

I. 80% of n = 32

J. 3% of x dollars = $.30

Pc3. Percentage. Solve for the missing numbers. Use dollar signs and decimal points where needed.

A. 5% of $☐ = $.25   B. 23% of x = 3.45   C. 8% of n dollars = $3.60

D. 16% of ☐ = 1.44   E. 7% of $125 =   F. 10% of $☐ = $350

G. 9% of x dollars = $.315   H. 5% of y dollars = $.76$\frac{1}{4}$ I. 15% of $9 =

J. 13% of z dollars = $2.47

Pd1. Percentage. Find the percentage.

A. ?% of 16 = 12

B. ☐% of 8 = 4

C. x% of 90 = $22\frac{1}{2}$

D. a% of 26 = 13

E. ?% of 9 = 3

F. ?% of 3 = 9

G. ☐% of 100 = 10

H. ☐% of $5 = $.25

I. What % of 39 is 13?

J. What percent of 25 is $6\frac{1}{4}$?

Pd2. Percentage. Find the percentage.

A. What % of $10 is $10?

B. What % of $100 is $125?

C. What % of 50 cents is $12\frac{1}{2}$ cents?

D. What % of 34 is 17?

E. What % of 78 is 13?

F. ☐% of 9 = 1.44

G. ?% of 125 = 21.25

H. x% of 188 = 21.24

I. ☐% of $10 = $.40

J. What % of $10 is $.25?

Pe1. Percentage. Do these problems by changing your percent figure to a fraction before solving.

A. $12\frac{1}{2}$% of 100 = $\frac{1}{8} \times \frac{100}{1} = \frac{100}{8} =$ B. $37\frac{1}{2}$% of 125 =   C. 50% of 91 =

D. $66\frac{2}{3}$% of 99 =   E. $62\frac{1}{2}$% of 64 =   F. $33\frac{1}{3}$% of 66 =   G. $16\frac{2}{3}$% of 132 =

H. $6\frac{1}{4}$% of 96 =   I. $3\frac{1}{8}$% of 128 =   J. $18\frac{3}{4}$% of 48 =

RATIO AND PERCENTAGE

Pf. Percentage. Solve by using the ratio method. (See page 76 if necessary.)

A. 25% of 16 =

B. 70% of 70 =

C. 50% of 90 =

D. 5% of $25 =

E. 6% of 66 =

F. 4% of $132 =

G. 15% of 50 =

H. 12% of $144 =

I. 9% of 300 =

J. 3% of $27 =

Pg1. Percent. Use any of the methods you have learned to solve the following problems.

How much tax do you pay on

A. $1.00 at the rate of 1%?

B. $5.00 at the rate of 2%?

C. $15.00 at the rate of 15%?

D. $20.00 at the rate of 5%?

E. $50.00 at the rate of 6%?

F. $75.00 at the rate of 8%?

G. $16.00 at the rate of $4\frac{1}{4}$%?

H. $36.00 at the rate of $16\frac{2}{3}$%?

I. $300.00 at the rate of 18%?

J. $10.00 at the rate of $2\frac{1}{2}$%?

Pg2. Percentage. Solve the problem below using the method you feel is best for you.

You buy an appliance that costs $360.00 plus 5% state sales tax. You agree to pay for the appliance in twelve equal monthly payments and pay all the sales tax with the first payment. The next eleven payments will have added interest at the rate of $1\frac{1}{2}\%$ a month on the unpaid balance. Make a chart like the one following and figure the total cost of your appliance.

Balance	Payment Number	Amount	Interest 1½% mo.	Sales Tax	Total Payment
$360|00	1		0|00		
	2				
	3				
	4				
	5				
	6				
	7				
	8				
	9				
	10				
	11				
	12				

TOTAL PAID_____

TOTAL INTEREST_____

Note: This problem clearly states that interest is paid on the unpaid balance. However, despite the truth in lending law, many credit granting groups are figuring interest on the balance *before* deducting payments and credits. This practice can raise your interest payment several percentage points. If you buy on credit, you would be wise to check carefully and determine exactly how your interest payment is being calculated.

WEEKLY TESTS I-XL

Following are forty weekly tests to be taken. Check each test before you hand it in for grading. Discuss with your teacher any problem area you find. This is important. If the teacher is too buy to discuss individual problems at that moment, wait until he/she can talk to you, or make an appointment to see him/her at another time, or ask someone else to help you. Don't repeat your mistakes. Try a new or different approach. Remember, learning the correct procedure puts money in your pocket.

WEEKLY TEST I

Add:

A. 324
659
378
107
966

B. $.15
+ .23

C. $ 1.29
3.97
.67
21.89

D. $.09 + $2.39 + $3.47 + $.27 =

E. $100 + $89 + $3.24 + $.09 =

WEEKLY TEST II

Add:

A. 2642
1857
3964
7283
4596
2862
7348
2587
4156
1329

B. $87.43
29.56
85.42
16.21
43.29
56.85
25.13
76.24
13.48
96.25

C. 7358
2519
1642
3725
8145
3729
1682
7326
2649
2834

D. 4519
6283
2846
3795
7642
3592
2841
7653
2164
9256

E. 9
56
374
2789
400
302
17
4
2001
566

WEEKLY TEST III

A. $15 + $3.19 + $2.08 + $.37 =

B. 456
− 321

C. 789
− 399

D. 8009
− 7777

E. $3.21
− .98

F. From $8 take 21 cents.

G. Take $.13 from $13.

H. 307
− 195

I. 4678
− 3989

J. $300.00
− 1.09

WEEKLY TEST IV

Add across and down. Prove your work.

$245.60	$117.26	$ 15.26	$ 6.25	E.
306.45	345.96	130.40	14.56	F.
129.52	475.58	75.93	7.38	G.
436.20	100.25	48.54	124.07	H.
493.18	326.45	9.65	56.35	I.
612.90	98.32	203.72	9.02	J.
A.	B.	C.	D.	

WEEKLY TEST V

A. 3 + 27 + 9 + 103 =   B. $100 + $3.27 + $16.98 + $5.49 =   C. 3093 + 2789 =

D. 15002   E. 8080   F. 234   G. 1023   H. 204   I. 908   J. 457609
   − 7083   − 7979   ×2   ×3   ×3   ×2   ×2

WEEKLY TEST VI

A. 324   B. 897   C. 5060   D. 736   E. 8007   F. 234   G. 898
   ×7   ×4   ×8   ×6   ×7   ×12   ×10

H. $3 + $.03 + $.09 =   I. 8030   J. 14235
                           −948   ×9

WEEKLY TEST VII

A. 149   B. $15.15   C. 9807   D. From 400 dollars take 44 cents.   E. 303   F. 123
   419   23.23   −9616   ×7   ×30
   914   49.49
   491   57.75
            98.89
            65.06

G. 303   H. 9008   I. 603   J. 6789
   ×33   ×9   ×7   ×9

WEEKLY TEST VIII

A. $3.21   B. 3)966   C. 4)1616   D. 3536   E. 2563   F. 303   G. 1241
    ×8   −2563   ×203   −98   ×92

H. 6677   I. 8989   J. 12)396
   ×8   ×76

WEEKLY TEST IX

A. 303 + 99 + 8 =   B. Take 16 away from 73.   C. 123 × 321 =   D. 4812 ÷ 4 =

E. 3603   F. 2)80604   G. 391   H. 8067   I. 7)1515   J. 13)39
   ×234   ×46   ×77

WEEKLY TEST X

A. 321   B. 30903   C. 7654   D. 13)3978   E. 29)60   F. 9)10890   G. 14)2926
    457   −9894   ×607
    689
    976
    453
    106

H. 30)61   I. 17)3468   J. 12345679
                                ×   63

WEEKLY TEST XI

A. 309   B. 807   C. 1313   D. 19)5871   E. $15.14   F. $3.02
   ×19   ×807   ×71   ×9   − .87

Reduce:

G. $\frac{9}{12} =$ H. $\frac{6}{8} =$

Raise to higher terms:

I. $\frac{1}{2} = \frac{}{8}$ J. $\frac{3}{4} = \frac{}{24}$

WEEKLY TEST XII

A. $3.02 + $9.98 + $8.76 =   B. 3030 − 999 =   C. 306 × 202 =   D. 4692 ÷ 23 =

Reduce all sums to lowest terms when possible.

E. $\frac{1}{3} + \frac{1}{3} =$ F. $\frac{2}{6} + \frac{1}{6} =$ G. $\frac{1}{8} + \frac{3}{8} + \frac{2}{8} =$ H. $\frac{1}{7} + \frac{2}{7} + \frac{4}{7} =$ I. $\frac{5}{16} + \frac{9}{16} =$

J. $\frac{3}{12} + \frac{4}{12} + \frac{2}{12} =$

WEEKLY TEST XIII

Reduce to lowest terms:

A. $\frac{13}{39} =$ B. $\frac{3}{15} =$ C. $\frac{23}{69} =$ D. $\frac{4}{12} =$ E. $\frac{18}{27} =$ F. $\frac{50}{100} =$

Continued on next page.

Subtract: (Reduce answers, if possible.)

G. $\frac{2}{3} - \frac{1}{3} =$ H. $\frac{3}{4} - \frac{1}{4} =$ I. $\frac{9}{10} - \frac{4}{10} =$ J. $\frac{13}{16} - \frac{7}{16} =$ K. $\frac{10}{15} - \frac{5}{15}$ L. $\frac{9}{12} - \frac{3}{12}$

M. $\frac{7}{8} - \frac{3}{8}$ N. $\frac{5}{6} - \frac{3}{6}$ O. $\frac{6}{7} - \frac{2}{7}$

WEEKLY TEST XIV

Follow Directions!

A. Subtract 28643796 from 83264972 and multiply your answer by 64. Divide the product by 3 and write the quotient on your paper.

B. From 46,298,725 subtract 19,327,367 and multiply the remainder by 32. Divide the product by 4 and write the quotient on your paper.

C. $12 \overline{)122436}$ D. 12345679×45

WEEKLY TEST XV

A. 327
    72
    918
    3
    406

B. 308095
    −117766

C. 3579
    ×88

D. $19 \overline{)57057}$

Express as Mixed Numbers. Reduce fractions to lowest terms.

E. $\frac{9}{5} =$ F. $\frac{11}{4} =$ G. $\frac{16}{14} =$ H. $\frac{24}{10} =$ I. $\frac{35}{30} =$ J. $\frac{48}{7} =$

Change to Improper Fractions.

K. $1\frac{2}{3} = \frac{}{3}$ L. $2\frac{7}{9} = \frac{}{}$ M. $3\frac{9}{10} =$ N. $5\frac{1}{2} =$ O. $1\frac{6}{15} =$

Check: Make sure that you followed directions.

WEEKLY TEST XVI

A. $\frac{1}{4} + \frac{1}{4} =$ B. $\frac{1}{3} + \frac{1}{3} =$ C. $\frac{1}{2} + \frac{1}{2} =$ D. $\frac{1}{2} + \frac{1}{3} =$ E. $\frac{1}{2} + \frac{1}{4} =$ F. $\frac{1}{4} + \frac{1}{4} + \frac{1}{4} =$

G. $\frac{1}{2} + \frac{1}{3} + \frac{1}{4} =$ H. $\frac{1}{2} + \frac{1}{3} + \frac{1}{6} =$ I. $\frac{5}{6} - \frac{2}{6} =$

J. Your lunch costs $1.85. You pay for it with a $5.00 bill. How much change should you get back from the cashier?

Check: Are fractions simplified in your answers?

WEEKLY TEST XVII

Reduce fractions to lowest terms.

A. $\dfrac{9}{10} + \dfrac{3}{5} =$ B. $\dfrac{5}{8} + \dfrac{1}{4} + \dfrac{1}{2} =$ C. $\dfrac{5}{8} - \dfrac{1}{4} =$ D. $\dfrac{14}{16} - \dfrac{8}{16} =$ E. $\dfrac{2}{3} - \dfrac{1}{6} =$

F. $\dfrac{1}{2} + \dfrac{3}{4} + \dfrac{7}{8} =$ G. $\dfrac{8}{9} - \dfrac{2}{3} =$ H. $\dfrac{9}{14} - \dfrac{2}{7} =$ I. $\dfrac{45}{46} - \dfrac{14}{23} =$ J. $\dfrac{1}{2} \times \dfrac{1}{3} =$

WEEKLY TEST XVIII

A. $15 + $15.15 + $30.95 =   B. Take $.79 from $10.   C. Multiply 2468 by $.15.

D. The FEAST Club made $9.25 selling bread at $.25 a loaf. How many loaves of bread did the club members sell?

E. $\dfrac{3}{4} + \dfrac{3}{8} =$ F. $\dfrac{3}{4} - \dfrac{3}{8} =$ G. $\dfrac{2}{3} \times \dfrac{2}{3} =$ H. $\dfrac{3}{4} \times \dfrac{2}{3} =$ I. $\dfrac{3}{4} \times \dfrac{3}{8} =$ J. $1\dfrac{1}{2} + \dfrac{1}{4} =$

WEEKLY TEST XIX

A. $\dfrac{1}{2} \times \dfrac{1}{5} =$ B. $\dfrac{1}{2} + \dfrac{1}{3} + \dfrac{1}{4} =$ C. $\dfrac{5}{6} - \dfrac{1}{2} =$ D. $3 - \dfrac{3}{8} =$ E. $3 - \dfrac{5}{8} =$ F. $1 - \dfrac{1}{4} =$ G. $1\dfrac{1}{2} + 2\dfrac{3}{4} + 3\dfrac{5}{8} =$

H. $15\dfrac{5}{8} - 12\dfrac{3}{8} =$ I. $10\dfrac{3}{4} \times 4 =$ J. $101\dfrac{9}{10} + 36\dfrac{1}{2} + 391\dfrac{2}{5} =$ K. $2001\dfrac{37}{39} - 999\dfrac{3}{13} =$ L. $409 \times 708 =$

WEEKLY TEST XX

A. $15\dfrac{5}{8} + 6\dfrac{5}{6} + 17\dfrac{1}{24} + 13\dfrac{7}{12} =$ B. $109\dfrac{13}{36} - 99\dfrac{1}{12} =$ C. $765\dfrac{3}{8} \times 48 =$ D. $36 \times 7\dfrac{2}{9} =$ E. $48 \div 13 =$ F. $\dfrac{20}{6} =$ G. $\dfrac{1}{2} \div \dfrac{1}{3} =$

H. $\dfrac{9}{10} \times \dfrac{5}{6} =$ I. $2\dfrac{7}{8} \times \dfrac{3}{4} =$ J. $\dfrac{3}{4}$ of $16 =$

WEEKLY TEST XXI

A. $\dfrac{1}{2}$ of $10 =   B. $\dfrac{7}{10} \div \dfrac{1}{3} =$ C. $\dfrac{3}{4}$ of $88 =$ D. $\dfrac{8}{9} \times \dfrac{2}{3} =$ E. $3\dfrac{3}{4} \times 5\dfrac{5}{6} =$

F. $5\dfrac{3}{5} \div 1\dfrac{1}{2} =$

Continued on next page.

Express as decimal fractions.

G. $\frac{1}{4} =$ H. $\frac{1}{2} =$ I. $\frac{1}{8} =$ J. $\frac{5}{10} =$ K. $\frac{3}{4} =$ L. $\frac{1}{3} =$ M. $\frac{2}{3} =$ N. $\frac{1}{5} =$

O. $\frac{1}{6} =$

WEEKLY TEST XXII

A. $103.33	B. $900.00	C. $144.27	D. $12\overline{)\$36.36}$	E. .2	F. $.23	G. $.78
2.98	− 89.89	× 67		.3	.34	× 34
15.13				.4	.45	
300.30						

Express the following decimal fractions as proper fractions.

H. $.33\frac{1}{3} =$ I. $.66\frac{2}{3} =$ J. $.16\frac{2}{3} =$

WEEKLY TEST XXIII

A. .3 + .6 + .7 + .9 =   B. 1.2   C. .9 − .4 =   D. 1.2   E. 16.7   F. 8.8
                           .6   − .9   − 10.8   − 1.8
                           3.4
                           5.5

G. 3.03   H. 8.07   I. 3.2 + 4.9 + .05 + 39.76 =   J. From 13 take .013.
  − 1.01   − 3.09

WEEKLY TEST XXIV

A. 3.3 + 4.44 + 5.555 =   B. Subtract .05 from 5.   C. .6 × .6 =   D. .2 × .3 =

E. 3.24   F. 6.67   G. 6.07   H. 982   I. 57.6   J. 8.09   K. 124.35   L. .1357
  × 5   × .6   × .07   × .37   × .28   × 32   × .67   × .6

WEEKLY TEST XXV

A. $1\frac{1}{2} \times 1\frac{1}{2} =$ B. 2.6 × 1.3 =   C. $\frac{3}{4} \div \frac{4}{7} =$ D. 30 + .27 =   E. $1\frac{3}{7} \div 2\frac{2}{9} =$

F. .6 ÷ 2 =   G. $.2\overline{)6}$ H. 708.97   I. $3.3\overline{)9.9}$ J. $.07\overline{).4949}$ K. $.007\overline{).9898}$
                                       × 1.4

L. $15\overline{)45.9}$

WEEKLY TEST XXVI

A. 3343
7764
8890
4760
1539

B. $155\overline{)31155}$

C. 8000
− 976

D. .65
× .65

E. $\frac{3}{8} \times \frac{3}{8} =$

F. .375
× .375

G. $5^3 = 5 \cdot 5 \cdot 5 =$

H. $.3 \times .3 \times .3 =$

I. $310\overline{)62.310}$

J. $\frac{39}{13} =$

WEEKLY TEST XXVII

A. $.11\overline{)6765}$

B. $.101\overline{).6666}$

C. $.234\overline{).57096}$

D. $1.25\overline{)153.75}$

E. $\frac{7}{8} \div \frac{3}{8} =$

F. $.375\overline{).875}$

G. $.8199\overline{).57393}$

H. $.004\overline{)4}$

I. $.3\overline{)36}$

J. $.09\overline{)81}$

WEEKLY TEST XXVIII

A. Take .003 from 1.2.

B. .007
× .006

C. 1.4
× 1.4

D. $3.91\overline{)15.2881}$

E. $3.91 + .007 + 1.4 + 12 =$

Express the following quotients correct to the nearest hundredth.

F. $3\overline{)2}$ G. $7\overline{)4}$ H. $9\overline{)8}$ I. $13\overline{)12}$ J. $15\overline{)9}$ K. $23\overline{)20}$ L. $17\overline{)16}$

WEEKLY TEST XXIX

Express the following correct to the nearest cent.

A. $1.097 = B. $.33$\frac{1}{3}$ = C. $3.495 = D. $15.157 = E. $6.034 =

Express the following correct to the nearest dollar.

F. $6.57 = G. $7.98 = H. $123.33 = I. $57.11 = J. $33.66 =

Express the following as percents.

K. .03 = L. .25 = M. .0325 = N. 1 = O. $\frac{1}{4}$ = P. $\frac{1}{3}$ = Q. .125 =

R. $\frac{3}{8}$ = S. $\frac{5}{10}$ = T. $\frac{7}{100}$ =

Express the following percents as common fractions.

U. 25% = V. 33$\frac{1}{3}$% = W. 12$\frac{1}{2}$% = X. 66$\frac{2}{3}$% = Y. 20% =

WEEKLY TEST XXX

A. 25% of $8 =   Solve B for "x."   B. $\frac{10}{100} = \frac{x}{20}$ C. 5% of what number = 3?

D. What percent of 16 is 4?   E. 33$\frac{1}{3}$% of 66 =   F. $\frac{1}{8}$ of 50 =

Continued on next page.

G. $7\frac{2}{3} \div 11\frac{7}{8} =$ H. $107\frac{1}{2}$
$\times 21\frac{1}{3}$ I. How many quarter pound patties can be made from 121 pounds of hamburger?

J. Salt costs 16¢ a pound. One tablespoon of salt weighs $\frac{1}{2}$ ounce. How much does one tablespoon of salt cost?

WEEKLY TEST XXXIX

A. 25% of $25 =   B. $33\frac{1}{3}$% of $198 =   C. $12\frac{1}{2}$% of $\square = 8$ D. 3 lbs. 7 oz. + 15 oz. =

E. At 15¢ an oz. how much do 3 lbs. cost?   F. 8 lbs. 5 oz.
$- 3$ lbs. 9 oz.   G. 3 lbs. 7 oz.
$\times 4$

H. $3\overline{)15 \text{ lbs. 6 oz.}}$ I. $3\frac{1}{2} \times 6 =$ J. $15\frac{1}{2} \div 3 =$ K. $325^2 =$ L. $9^2 + 8^2 + 3^2 =$

WEEKLY TEST XL

A. Multiply 6.047 by 23.   B. Subtract 23.03 from 30.   C. Add 15.15, 3.03, 14.1, .09, 31.

D. Multiply $\frac{2}{3}$ by $\frac{3}{4}$ and divide the result by $\frac{5}{6}$.   E. $14.14 \div 1.4 =$ F. $6\overline{)37 \text{ lbs. 14 oz.}}$

G. 304.56
$\times .78$ H. At 96¢ a gallon, how much does 1 cup cost?

I. $1\frac{1}{4}$ oz. of ground ginger costs 35¢. At this rate, how much do 2 oz. cost?

J. Criss starts washing dishes at $65.00 a week. A month later he gets a 4% increase. Because of business problems, Criss gets a 4% decrease 2 months later. How much more or less is Criss paid after the cut than he got when he started? (Be sure to label your answer, "more" or "less.")

BUSINESS SITUATIONS INVOLVING MATHEMATICS

SANITATION

Some germs are useful to humans, and some germs do not affect us, but many germs are very dangerous to people, especially in food. For example, germs carry disease and cause food poisoning.

Germs are also called *bacteria*. They are alive but are so small that we cannot see them without a microscope. And they are everywhere, particularly on people—in our breath, in discharges from our bodies, on our skin, in our hair, on our clothing. Because our hands are always touching something, they probably pick up more bacteria than any other part of our body.

Bacteria multiply quickly at medium temperatures in the presence of food, oxygen, and moisture. They are killed by high temperatures, and their growth is slowed down by freezing.

Since there are bacteria everywhere, there are always some in food, even in good, clean food. Our job in the foods business is to prevent disease-carrying bacteria from entering food and to prevent bacteria from growing and multiplying in food to the point where it becomes dangerous to the people who eat it. We do this in two ways basically:

1. Keep everything around food clean.

2. Keep food in a condition so that the few harmless germs in it do not multiply and become dangerous.

Here are some rules that *everyone in the foods business* should observe to prevent sickness and food poisoning:

PERSONAL HYGIENE

1. If you are sick, stay home from work.

2. Bathe with soap every day.

3. Wear clean clothes.

4. Treat and cover open sores.

5. Wash your hands and arms whenever they get dirty.

6. *Always* wash your hands after using the toilet or a handkerchief.

7. Wash your hands after handling garbage or poisons.

8. Don't cough or sneeze near food or dishes.

9. Don't touch food with your hands—use the proper implements or disposable foodservice gloves.

10. Keep fingers out of dishes and glasses and off the serving ends of knives, forks, and spoons.

11. Keep fingernails trimmed and clean.
12. Wear a clean apron or uniform.
13. Keep your hair neat. Wear a head band, net, or cap.

FOOD SANITATION

14. Store food in clean, dry places.
15. Store foods that can spoil in a refrigerator.
16. Remove food from opened cans and place in porcelain or glass container for storage.
17. Keep dirty dishes away from food.
18. Keep foods covered.
19. Recook any leftover food.
20. Never re-use food left by customer.
21. Refrigerator should be at 45° F. or lower.
22. Refrigerator should be clean.
23. Place food in refrigerator with space for air circulation.
24. Keep hot foods hot (over 140° F.) and cold foods cold (under 40° F.). Food should not be in between these temperatures more than 4 hours total time.
25. Cover food waiting to be served.
26. Keep poisons away from food.

DISHWASHING

27. Keep dirty dishes away from clean dishes.
28. Use hot water—110° F. by hand, 140° F. by machine.
29. Scrape and pre-rinse dishes before washing (Necessary to remove drying egg, etc.)
30. Change water frequently.
31. Use a clean tray for sanitizing dishes.
32. Sanitize—bathe dishes at 170° F. for $\frac{1}{2}$ minute or use a chlorine bath.
33. Store dishes on clean shelves, high above floor and dust.
34. Store glasses and cups upside down.
35. Don't use dish towels—dry by draining or heat.
36. Keep dippers, scoops, etc., used for frozen foods under *running* water between washings.

FOOD AREAS

37. Have good ventilation.
38. Keep out flies, rodents, etc.
39. Throw out waste.

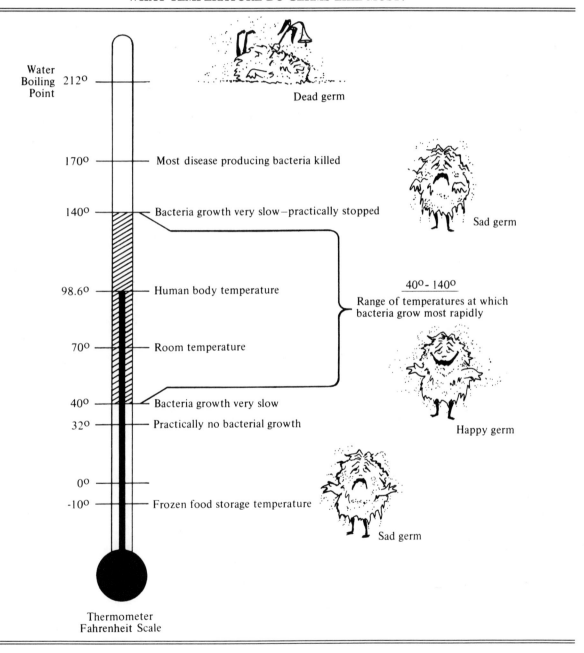

113

HOW FAST DO GERMS MULTIPLY?

For bacteria (germs) to reproduce they need:

    food

    moisture

    oxygen

    temperature from 40° F. to 140° F.

Bacteria reproduce by dividing in half.
    Under ideal conditions they divide every 20 minutes.
    One bacterium becomes two, two bacteria divide into four, four divide in half and become eight.

PROBLEM:

Starting with one bacterium at noon and multiplying by two every 20 minutes, how many bacteria would the one become

    at 4:00 o'clock?

    at 6:00 o'clock?

    at 8:00 o'clock?

Time	Bacteria
12:00 noon	1
12:20	2
12:40	4
1:00	8
1:20	16
1:40	32
2:00	64
2:20	?
2:40	?
3:00	?
3:20	?
3:40	?
4:00	?

QUESTIONS

1. What is another word for germs?

2. Are all germs harmful to humans?

3. Can you see germs with the naked eye?

4. What four things do bacteria need to multiply quickly?
5. What kills bacteria?
6. What effect does freezing have on bacteria?
7. What are two basic things people in the foods business do to keep bacteria from multiplying?
8. Name two times when you should always wash your hands.
9. How hot should hot foods be kept for good food sanitation?
10. How cold should cold foods be kept for good food sanitation?
11. How long is it safe to keep food at medium temperatures? (For example, room temperature?)
12. How should glasses and cups be stored?
13. Should dishtowels be used for drying dishes?
14. Give two sanitation rules for left-over food.
15. Give two sanitation rules for food areas.

WAGE SCALES

When people talk of *wages,* they usually mean pay for work that is figured on an hourly or daily basis. *Salary* means a fixed amount of pay for a longer period of time, for example, a weekly or a monthly salary.

COLLECTIVE BARGAINING AGREEMENTS

Wage scales and working conditions in hotels and restaurants that hire union employees are determined by a collective bargaining agreement. The officers of the union and representatives of an association of restaurant and/or hotel owners work out the details. Once these people reach an agreement, it is usually approved by the other union members and the individual owners and then signed into a contract, or collective bargaining agreement, for a period of time, such as two or three years.

NONTIPPED EMPLOYEES AND TIPPED EMPLOYEES

There are two classifications of employees for hotel and restaurant wage scales: nontipped employees, such as cooks, pantry persons, and other people working in the back of the house; and tipped employees, such as waiters, waitresses, bartenders, and others working in the front of the house. These people in the front of the house are really paid partly by their employer and partly by the guests they serve. Therefore, the wages they receive from the employer are less than the wages received by nontipped employees.

The wages agreed upon are minimum rates of pay. In many places of business a good cook, or waiter, or bartender may be paid more than the minimum. In other places, where the employees do not belong to the union, the pay might be less than the union wage scale in that area. However, employers must obey federal and state laws regarding minimum wages.

WHAT IS A WORK DAY?

A full day's work for a culinary worker normally consists of 7½ hours of actual work over a period of 8 hours. The employee is paid for 8 hours, but he or she eats a meal during ½ hour of this time.

In restaurants, people sometimes work a split shift or a short shift. On split shift, a person works for 3 or 4 hours, is off for 3 or 4 hours, and then comes back to work the remaining 3 or 4 hours of his or her shift. A short shift is where a person works for less than a full shift in one day, such as for 3 or 4 hours.

For working a split shift or a short shift, some employer-union agreements call for a higher rate of pay than for a regular straight shift. This is called premium pay. The employees are paid a premium wage for working inconvenient hours.

TRAINING AND APPRENTICESHIP PROGRAMS

Many collective bargaining agreements have special rules and wage scales for apprentices. An apprentice is an employee who is learning on the job. The usual arrangement calls for the employee to work alongside experienced employees and, in addition, attend instructional classes each week. After a period of months or years as an apprentice, the employee becomes a "journeyman." A journeyman is an experienced, skilled worker, who is entitled to full pay and working conditions. The apprentices are usually paid a percentage of the journeyman employees' wages, and the percentage increases as the apprentice learns and gains experience.

PROBLEMS

In the following problems round off all answers to dollars and cents. Some of the daily wage scales that were in effect in a typical employer-union collective bargaining agreement during 1987 are given below.

I. Figure out the straight shift pay per hour for each job.

II. The pay for a split shift is 15% more than for a straight shift. Figure the total day's pay for a split shift for each job. To get the answers simply multiply the Straight-Shift 8 Hours Pay by 115%. Remember to convert the 115% to the decimal 1.15 before you multiply.

	Straight Shift 8 Hours	*I. Straight Shift Per Hour*	*II. Split Shift 8 Hours Work in 12 Hours*
1. Chef or Head Cook	$94.65	_____	_____
2. Second Cook	78.15	_____	_____
3. All Other Cooks	72.53	_____	_____
4. Cashier	54.90	_____	_____
5. Garde Manger*	68.03	_____	_____
6. Waiter/Waitress	37.50	_____	_____
7. Busperson	43.50	_____	_____
8. Supply & Dish Up	49.43	_____	_____
9. Doorperson	53.70	_____	_____
10. Service Bartender	76.73	_____	_____

*A *garde manger* is a person who has charge of the cold meat department. He oversees (1) breading of seafood, poultry, etc.; (2) preparation of salad dressings and other cold sauces; (3) preparation of meat, fish, and seafood salads; (4) preparation and decoration of all cold food for buffet service; and (5) the making of appetizers, canapes, and sandwiches.

III. Figure out the apprentice daily pay for the following two groups:
   A. Union cooks earn $64.28 per day. The apprenticeship rates are:
      1. 1st 12 months 55%. Pay is _____
      2. 2d 12 months 70%. Pay is _____
      3. 3d 12 months 90%. Pay is _____
   B. Union bartenders earn $73.05 per day. The apprenticeship rates are:
      1. 1st 6 months 65%. Pay is _____
      2. 2d 6 months 80%. Pay is _____
      3. 3d 6 months 90%. Pay is _____

QUESTIONS

1. Define the following two terms:
   a. Wages
   b. Salary

2. What is a collective bargaining agreement?

3. Do cooks, waiters, bartenders, etc., ever receive more than the minimum union wage?

4. What is a split shift?

5. What is a short shift?

6. What is a garde manger?

7. What are gratuities?

8. What is an apprenticeship program?

INFLATION AND THE CONSUMER PRICE INDEX

Inflation can cause businesses to go broke and people to go hungry. What is inflation? Inflation is simply higher prices. What is the Consumer Price Index (also known as the CPI)? It is a group of figures that show how much higher prices are. If prices were to go down, that would be called deflation, and the Consumer Price Index figures would then show how much lower prices are.

WHAT INFLATION MEANS TO YOU

Inflation means that you have to pay more money today than you did in the past to buy the same thing. A family in San Francisco bought a compact station wagon in 1979 for a total cost of $5,461. In 1987 they bought a new wagon, same make and model, and the total cost was $10,393. During those 8 years, the price of the car almost doubled. That is inflation.

INFLATION HURTS

Inflation hurts almost everyone. For example, in building a new restaurant, the owners might plan to spend $350,000 for construction and other expenses. During the six months it takes from the planning stage to the actual opening of the restaurant, inflation can bring the total cost up to $385,000 or more. This means the owners must come up with an additional $35,000.

It also means that the owners will have to charge $12 or $13 for dinners instead of the $10 they had originally planned on charging. But will enough customers pay that higher price for dinner for the restaurant to succeed? Some difficult decisions have to be made by businesspeople because of inflation.

HOW THE CONSUMER PRICE INDEX IS FIGURED—THE "MARKET BASKET"

The CPI is based on a sample of all the goods and services that people spend money for in day-to-day living. The items selected for the sampling of prices and fees are called the "market basket" of goods and services. The CPI compares the market basket costs this month with the same market basket costs a month, a year, or 10 or 20 years ago.

The market basket contains hundreds of items. They are listed in seven major groups. The groups and examples of categories within each group are as follows:

Food (cookies, cereals, cheese, coffee, chicken)
Housing (residential rent, fuel oil, soaps and detergents, local telephone service)
Apparel and its upkeep (men's shirts, women's dresses, jewelry)
Transportation (airline fares, new and used cars, gasoline, car insurance)
Medical care (prescription drugs, eye care, physicians' services, hospital rooms)
Entertainment (newspapers, sports vehicles, toys, musical instruments, admissions)
Other items (haircuts, college tuition, bank fees)

There are individual indexes for the categories listed above, which are combined into one index. There are indexes for various areas of the United States, which are combined into one national average.

The base year is 1967. The prices of the market basket in 1967 are used as 100%, or simply 100. The CPI measures the change in prices since the base year. A CPI of 331.1 at the end of 1986 means that prices at that time were 331.1% of the prices for a similar market basket in 1967. Prices rose 231.1% from 1967 to December 31, 1986.

Each month trained employees of the Bureau of Labor Statistics of the Department of Labor visit, telephone, and send questionnaires to thousands of retail stores, service establishments, rental units, doctors' offices, etc., all over the United States in order to obtain price information on each of the many items in the market basket. The results of this sampling of prices are combined by the Bureau into the latest national CPI and announced to the public.

Detailed data on individual items or areas of the country can be obtained either by telephoning the Bureau of Labor Statistics or going to its office.

HOW THE CPI IS USED—WHO NEEDS IT?

Almost all Americans are affected by the Consumer Price Index. It is used in three major ways:

1. *As an indicator of how the economy is doing.* It indicates how well the government is controlling inflation. The President, Congress, and the Federal Reserve Board watch the CPI when making decisions regarding government income and spending and banking policies. In addition, business executives, labor leaders, and other citizens use the CPI as a guide when making economic decisions.

2. *As an adjuster for other economic figures.* For example, economists calculate how much goods and services are produced by the entire United States each year. This is called the Gross National Product. The total cost of goods produced and services performed for the year might add up to 5% more than the previous year. But economists, knowing that the Consumer Price Index rose 5%, would realize that the country did not produce any more—it merely paid 5% more for the same production.

Similarly, if a restaurant owner has had to raise prices 5% to keep up with a 5% increase in costs, his or her sales for the year might show an increase of 5%; but the owner has not increased business at all. Even if his or her profits have increased 5%, the owner has not gained: because everything he or she buys with the profits now cost 5% more.

3. *As a guideline for setting wages and income payments.* More than 4 million workers are covered by collective bargaining agreements

that tie wages to the CPI. That means there is a contract between the unions and the employers which states that the wages paid will be adjusted up or down according to how much the CPI goes up or down. These are known as "escalator clauses." The word "escalate" means rise or increase.

About 38 million people receiving Social Security benefits and another 3.5 million people who receive military and Federal civil service pensions receive increases in their income payments when the CPI shows there is inflation.[1]

Some landlords and tenants have escalator clauses in their lease agreements, which mean rent increases when the CPI increases. Some alimony and child support agreements use escalator clauses and the CPI. Finally, in recent years the CPI has been used to adjust the Federal income tax rates to prevent increases in taxes being paid by taxpayers because of inflation.

THE INFLATION RECORD IN THE UNITED STATES

Over a period of years, especially since World War II ended in 1945, the United States has had some inflation of prices almost every year. For a few years, prices went up 1 to 3 percent each year over the previous year. In the years 1949 and 1955 prices actually went down a little; we had deflation. Then during the 1960s, 1970s and 1980s it was inflation every year, sometimes very big and serious.

If you look at the graph on page 123, you can see that from the base year 1967 to the end of 1972 the CPI went up about 4 to 6 percentage points a year. In 1973, however, prices began to increase rapidly. By December 1978 the CPI had reached 202.9. This means that prices in December 1978 averaged 202.9% of the prices in 1967, or a little more than double in 11 years. 1980 was the worst year for inflation in the United States in the past 25 years—the CPI went from 229.9 to 258.4, an increase of 28.5 points from January 1 to December 31.

These figures for 1980 can be looked at from two points of view:

1. The CPI of 258.4 at the end of 1980 means that prices were 258.4% of the average prices in 1967.

2. The increase of 28.5 points in the CPI in 1980 is approximately 12.4% for the year. The inflation *rate* was 12.4%. The rise of 28.5 CPI points is 12.4% of the 229.9 CPI at the beginning of the year.

While a 12% inflation rate for a year is too high for economic comfort, it does not compare with the extreme conditions in some other countries where the rate of inflation at times has been more than 100% per year! Inflation that high means that the country is in serious trouble and many people will not be able to pay for basic food, shelter, and clothing.

CAUSES OF INFLATION

What causes inflation? There are many opinions, and a complete explanation would be quite complicated. Basically, though, the amount of

---

[1] "Revising the Consumer Price Index," Report 734, U.S. Department of Labor Bureau of Labor Statistics, November 1985.

money available in the nation is increasing faster than the amount of goods and services available. When there are more dollars available to buy the same amount of goods and services, the prices for this market basket tend go up.

Labor contracts were mentioned above. An escalator clause helps workers keep up with inflation. Other people attempt to keep up by increasing the prices of the goods they sell, or the services they perform, raising rents, etc. While these increases help those individuals temporarily, they also add to the inflation for everyone; so it creates a vicious circle.

Retired people living on their pensions and lifelong savings suffer most from inflation. They have practically no way of increasing their income as prices increase; so they end up buying less and less with their fixed incomes. Social Security payments increase as the CPI goes up, but not every senior citizen is eligible for Social Security.

GOVERNMENT ATTEMPTS AT CONTROL

The government, knowing that many people suffer from the serious effects of inflation, tries to control it. One way to slow down inflation is to cut down the amount of money in circulation by the following methods:

1. Making it difficult to borrow from banks. This is done by the control of interest and credit requirements through the Federal Reserve System.

2. Cutting down government spending.

3. Raising taxes so that people have less money to spend.

4. Setting nationwide wage and price controls. This means it is against the law to raise wages or prices more than the government allows, say 5%, while this law is in effect. However, the law is unpopular with most Americans; so the government only puts it into effect in a real emergency and for a short period of time, such as 12 months.

Most people would agree, however, that the government alone cannot slow down inflation. It also takes the cooperation of businesspeople, labor leaders, and the public.

PROBLEMS

Listed below are some prices of goods and services in 1967, the year the government uses as the base year for calculating the Consumer Price Index. The prices in 1967 are considered to be 100%. Also

	Average Price 1967	Consumer Price Index—1986	Price in 1986
1. Hamburger	$ .59	287.3	
2. Entertainment tickets	2.50	292.2	
3. Rent, residential	112.50	286.0	
4. Restaurant meals	3.80	367.1	
5. New automobiles	2,850.00	232.2	
6. Boys' shirts	5.50	202.3	

CONSUMER PRICE INCREASES IN THE UNITED STATES
FROM 1967 TO 1987 (1967 = 100)

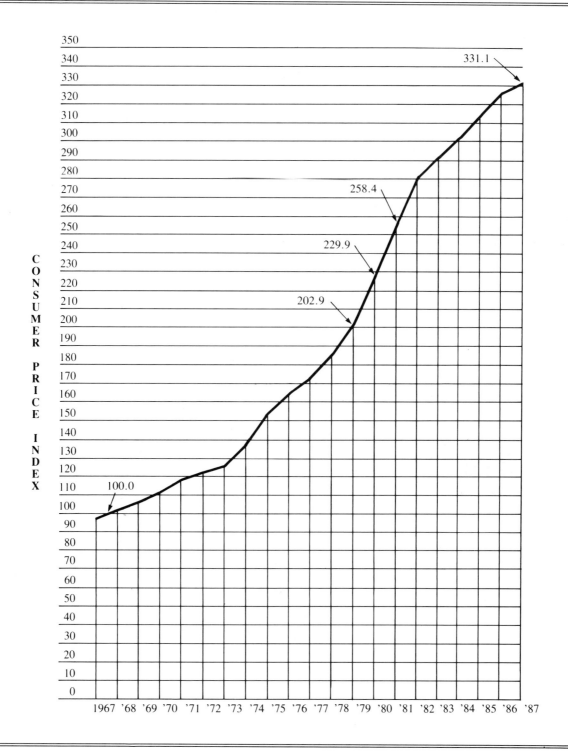

shown are the CPI figures for each item 19 years later, December 1986.

Find out what the prices were in 1986 by multiplying the 1967 prices by the 1986 Index. (Remember, the Consumer Price Index figures are actually percent figures. Therefore, they must be changed to decimals by moving the decimal point two places to the left before multiplying.)

In the table below are prices in 1967 and 1986. Find the Consumer Price Index for 1986 by dividing the 1986 price by the 1967 price.

Remember, index figures are actually percentages; therefore, you will move the decimal point in the answer 2 places to the right to convert it from a decimal number to a Consumer Price Index figure, as shown in the U.S. Bureau of Labor Statistics. (Round off your answers to the third place after the decimal point before you move the decimal point two places to the right.)

	Average Price 1967	Average Price 1986	Consumer Price Index—1986
7. Gasoline	.34	.89	
8. Men's suits	68.40	138.37	
9. Appliances, radio & TV	102.70	207.97	
10. Hospital rooms	48.00	279.84	
11. Refrigerators	224.00	453.60	
12. Restaurant meals	3.80	13.95	

QUESTIONS

1. What is inflation?
2. What is the Consumer Price Index?
3. What does inflation mean to you?
4. Who is hurt by inflation?
5. What is the "market basket," and how is it used in the Consumer Price Index?
6. How many items are in the market basket?
7. Is the CPI figured only on goods sold, or does it include fees for services performed by people such as repairers, doctors, and accountants?
8. What year is the base year?
9. Exactly, what does the CPI measure?
10. Who gathers information used to calculate the CPI?
11. Where can you find out the latest Consumer Price Index?
12. What are the three major ways the CPI is used?
13. According to the graph, what was the CPI approximately at the beginning and end of 1985?
14. What is the basic cause of inflation?
15. What group of people usually suffers most from inflation?

LABOR COSTS

In other chapters of this book you learn about some of the fringe benefits that employees in the hotel and restaurant business enjoy in addition to their wages and tips. They include such things as paid vacations, extra pay or time off on holidays, free meals, and discounts, as well as contributions to insurance plans, retirement plans, and social security. Some employers now make contributions to dental and visual health plans for employees.

For some things, such as social security taxes, both the employer and the employee make payments. Most of the benefits are paid by the employer, however, and the employer figures them as part of his or her labor costs and the expense of operating a business. The employer realizes that the cost of hiring an employee is not just the amount of wages to be paid; it is wages plus another 20 to 30 percent for fringe benefits.

When a hotel rents a room or a restaurant serves a meal to a guest, the managers know that they must charge enough to cover the cost of the fringe benefits for employees.

Employee benefits vary somewhat with different locations and different employers; however, some benefits are prescribed by law and others are now so common that an employee can expect to receive them from just about any employer in the hospitality industry.

Below are some of the typical costs that restaurant owners had for fringe benefits, in addition to regular wages, in 1987:

Taxes, insurance, vacations

Social security taxes (employer's share)	$ .0715	on each dollar of wage
State unemployment insurance (California)	.023	" " " " "
State employment training tax	.001	" " " " "
State disability insurance	.009	" " " " "
Federal unemployment insurance	.062	" " " " "
Workmen's compensation insurance	.0379	" " " " "
Vacation fund (2 weeks paid vacation)	.04	" " " " "
Total cost—taxes, insurance, vacations	$ .2444	(Round off to $.24)
	$ .24	(or 24%) on each dollar of wages paid

Other labor costs

Accident and sickness fund	$ 3.057	per day per employee
Dental plan	.784	" " " "
Allowance for uniforms	1.50	" " " "
Meals—3 per day	4.50	" " " "
Total other labor costs	$10.841	(Round off to $10.84) per day per employee

ACCOUNTING FOR MEALS

For accounting purposes, and particularly for income tax purposes, the employer adds $1.50 per meal to the wages each employee receives each day. Then the employer charges the employee $1.50 for each meal he or she eats while on the job. In this way the employee gets his or her meals free. The addition and the subtraction of the money for meals shows on the employee's paycheck stub and is reported to the government. The employees in restaurants and hotels do not pay income taxes on the value of the meals received while at work, but they do pay social security.

PROBLEMS

Figure the labor costs for a workday on each of the employees listed below.

Follow these steps for each employee:

a. Multiply the wages per day shown in column I by 24% to find the amounts for column II, the cost of taxes, insurance, and vacations that the employer must pay. Remember, a percent figure is converted to a decimal figure for multiplying by moving the decimal point two places to the left. Therefore, 24% = .24. Round off to pennies.

b. Add column I, Wages Per Day; column II, Cost of Taxes, Insurance, Vacation Per Day; and column III, the Other Employee Costs Per Day, to find column IV, Total Labor Costs Per Employee Per Day.

	I Wages Per Day	II Cost of Taxes, Insurance, Vacation Per Day	III Other Employee Costs Per Day	IV Total Labor Costs Per Employee Per Day
1. Chef or Head Cook	$94.65		$10.84	
2. Second Cook	78.15		10.84	
3. Night Cook	77.03		10.84	
4. Garde Manger	68.03		10.84	
5. Grillperson	72.53		10.84	
6. Sandwichperson	60.98		10.84	
7. Fast Foods Cook	56.63		10.84	
8. Griddle Cook	63.68		10.84	
9. Maitre d'	71.73		10.84	
10. Waiter Captain	62.48		10.84	
11. Waiter & Waitress	37.50		10.84	
12. Cashier	54.90		10.84	
13. Cafeteria Food Server	57.23		10.84	
14. Hostess	58.80		10.84	
15. Dishwasher	49.43		10.84	
16. Service Bartender	76.73		10.84	

	I Wages Per Day	II Cost of Taxes, Insurance, Vacation Per Day	III Other Employee Costs Per Day	IV Total Labor Costs Per Employee Per Day
17. Banquet Bartender	68.70		10.84	
18. Fountain Server	49.28		10.84	
19. Doorperson	53.70		10.84	
20. Parlor Maid	47.85		10.84	

WORKING CONDITIONS IN THE INDUSTRY

FRINGE BENEFITS

People working in the hospitality industry receive many *fringe benefits,* that is, things your employer gives you in addition to your salary.

Almost all types of workers get some kinds of fringe benefits, depending on where they work and what kind of work they do. If you work for an airline, or a steamship company, or a railroad, you get reduced rates when you travel. Most employees receive vacations with pay, payments to their social security account and their workmen's compensation account and other benefits.

People working in the hospitality industry get something that most other people do not get, however; and that is free meals while they are at work. In some hotel and motel jobs, they also get free lodging, or reduced rates for themselves and their families.

PROBLEMS

I. 1. In one hotel all employees receive one free meal each day. They select their meal from a buffet, which is sold to guests at the hotel for $9.95 per meal. During one month, John Williams, one of the waiters, had 23 free meals from the buffet. What was the value of the meals that he received free?

2. In another city a restaurant has an agreement with the union that they give each employee three free meals each day they work. The total cost for these three meals averages $8.65 per day. The cook worked twenty-four days during the month. What was the value of the food he ate during those twenty-four days?

3. A snack bar lets the counterpeople drink all the soft drinks they want on days that they work. The soft drinks sell for 65¢ each. In one week, one of the counter boys, Paul, had 13 drinks. What was the value of the free drinks that Paul drank?

4. In a drive-in restaurant, carhops are allowed one hamburger and one soft drink per day. The soft drinks sell for 58¢ and the hamburgers sell for $1.35. In one week, Erin, the carhop, had five drinks and five hamburgers. What was the total value of the food and drinks she got free?

5. In another restaurant, employees get a discount of 20% on meals purchased by them and their faimilies when they are off duty. How much would Harry, the fry cook, save on meals that cost a total of $87.35 during the month?

6. The manager of a large restaurant earns a salary of $2,500 per month, plus 5% of the income on all banquets the restaurant

serves. During the month of June, the receipts from banquets totaled $18,782.50. How much was the manager's commission on the banquets?

7. In an ice cream and candy store, the employees get a discount of 15% on everything they buy, including gift articles. In one month, Jimmy, the fountain attendant, bought candy and gifts that sell for a total of $37.35. How much was his 15% discount on this?

8. A motel has a novelty shop. Employees of the motel get a 20% discount on all purchases. During the month, Ella, the cashier, bought $38.95 worth of merchandise. How much did her 20% discount amount to?

9. At a ski resort, buspersons who work in the dining room get a free ticket on the ski lift each day, valued at $23.00, 1 free meal, valued at $3.25, and wages of $5.17 an hour. They work seven hours a day. What is the total value they receive each day, including wages, free meal, and free ski lift ticket?

10. The manager of a large hotel is paid a salary of $2,800 a month. In addition, he and his wife live in the hotel and get all their meals free. To pay for an apartment equal to their hotel accommodations would cost $1,200 a month. The meals they eat have an average value of $34 per day total for the two of them.
    a. Figuring an average of 30 days in a month, what is the total value of the meals they eat per month?
    b. What is the total value the manager is receiving per month in salary and food and lodging for him and his wife?

II. Many of the examples of fringe benefits given above required figuring percentage of dollars and cents. Here are some more problems to practice figuring percentage.

1. 25% × $13.00 =
2. 35% × $15.00 =
3. 9% × $29.95 =
4. 9½% × $82.63 =
5. 27% × $36.80 =
6. 27½% × $58.25 =
7. 37½% of $28.00 =
8. 33⅓% of $84.95 =
9. 22.5% of $67.50 =
10. 20.5% of $47.65 =
11. 18% × $55.00 =
12. 18½% × $65.00 =
13. 17.5% × $48.00 =
14. 60% of $27.50 =
15. 62.5% of $83.50 =
16. 24% of $98.25 =
17. 66.6% of $39.75 =
18. 32% of $47.35 =
19. 12.5% × $87.50 =
20. 7½% × 68.95 =

WORKMEN'S COMPENSATION INSURANCE

Like most people who work for wages, employees working in hotels, motels, and restaurants receive insurance payments if they are injured while at work. This insurance is called workmen's compensation insurance, and the cost of having the insurance is paid by the employer for the employees. In other words, the insurance is free to the workers.

The employer pays different rates for the insurance for different classifications of employees, such as cooks, waiters, cashiers, and front office clerks. The past record of safety of employees in that particular place of business also affects the rate paid. If the hotel or restaurant has a good safety record and not many workers have collected insurance for injuries, then the employer pays less for the insurance.

The cost to the employer is stated in terms of being so much for every $100 paid to the employees in wages or salaries; for example, $4.00 per each $100, or perhaps $4.25 per each $100 of wages paid.

One simple way to figure the expense, or total cost, is to divide the total wages by 100, and then multiply the result by the insurance rate. To divide any whole number or decimal by 100, you simply move the decimal point two places to the left. Therefore, if you wish to calculate the cost for insurance for six waiters who were paid a total of $4,525.00 for the month and the insurance rate for waiters is $4.15 per $100, you would figure it like this:

a. Divide the $4,525.00 by 100 by moving the decimal point two places to the left.

$4,525.00 = $45.25

b. Multiply that answer, $45.25, by the insurance rate, $4.15.

```
    $45.25
    ×4.15
    22625
     4525
    18100
  $187.7875
```

The answer then is $187.7875, which rounds off to $187.79. In other words, it costs the employer a total of $187.79 for workmen's compensation insurance for his six waiters for the month.

PROBLEMS

I. Figure the cost for workmen's compensation insurance for each of the following groups of employees for the month. (Round off answers to the nearest whole penny.)

1. Eight waiters were paid a total of $6,000.00. The insurance rate for waiters in this restaurant is $3.80 per $100.00 of wages paid.

2. Three cooks were paid a total of $4,110.00. The rate for cooks is $5.03 per $100.00.

3. Four buspersons were paid a total of $3,752.00. The rate for buspersons is $3.87 per $100.00.

4. The maitre d'hotel was paid $1,775.00 for the month. The insurance rate on him is $3.05 per $100.00.

5. The baker was paid $1,663.00 for the month. The rate for the baker is $4.83 per $100.00.

6. The two dishwashers were paid a total of $2,537.00 for the month. The rate for the dishwashers is $4.17 per $100.00.

7. The two cashiers were paid a total of $2,392.00 for the month. The rate for the cashiers is $2.1875 per $100.00.

8. The janitor was paid $1,550.50 for the month. The rate for the janitor is $3.736 per $100.00.

9. Four cocktail waitresses were paid $3,240.00 for the month. The insurance rate for the cocktail waitresses is $3.625 per $100.00.

10. Three bartenders were paid $4,743.75 for the month. The insurance rate for the bartenders is $3.09 per $100.00 of wages.

WHAT DOES AN EMPLOYEE COST YOU, MR. OWNER?

II. Here are some more problems to practice figuring the cost of workmen's compensation insurance—

1. An insurance rate of $1.83 per $100.00 on wages of $4,600.00.

2. A rate of $3.07 per $100.00 on wages of $3,185.00.

3. A rate of $2.41 per $100.00 on wages of $5,050.00.

4. A rate of $4.025 per $100.00 on wages of $2,695.00.

5. A rate of $1.89 per $100.00 on wages of $1,780.00.

6. A rate of $2.22 per $100.00 on wages of $3,649.00.

7. A rate of $1.963 per $100.00 on wages of $2,064.00.

8. A rate of $1.788 per $100.00 on wages of $1,904.50.

9. A rate of $2.005 per $100.00 on wages of $972.75.

10. A rate of $2.167 per $100.00 on wages of $8,934.25.

QUESTIONS

1. Name three fringe benefits that people in the hotel and restaurant business usually receive.

2. Who pays the employees social security taxes?

3. What is workmen's compensation insurance?

4. Who pays for workmen's compensation insurance?

5. Is the workmen's compensation insurance rate charged to the employer the same for all employees?

USING STANDARDS IN FOODSERVICE

Efficient restaurant people cook and bake with standard recipes. From these standard recipes they get standard yields. They serve standard portions in a style called a standard dish-up. The meaning of these words is very simple and the whole idea of using standards is just using common sense.

Standard Recipes

A standard recipe is a written formula (recipe) for producing a food item of a certain amount and quality; that is, you use exactly the same amount of ingredients and prepare the food in exactly the same way each time. If you also use ingredients that are similar in quality each time, then you will get the same results each time.

If all chefs use a standard recipe for chicken pie, no matter which chef is on duty and prepares the chicken pie, you will always get the same number of pies and they will always taste the same.

Standard Yields

The word "yield" means the actual amount of servable food you get from a standard recipe.

For example, you might start with twenty pounds of turkey for your recipe for turkey dinners; but by the time you get through preparing and cooking the turkey, you might have only twelve pounds of actual turkey meat that you can serve. You have lost eight pounds in carving, removing bones and other waste parts, shrinkage from cooking, and so forth.

The original weight of the turkey, twenty pounds, is the raw, or "as purchased," weight. The amount left for actual serving to the guests is the yield from the preparation and cooking.

If a standard recipe is followed every time, you should come out with a standard yield; that is, the same amount or weight of servable food, every time.

Standard Portions

Standard portions simply means serving the same amount of any particular item on the menu to each guest.

In one big chain restaurant in northern California the standard portions served on their turkey dinner consist of exactly:

2 oz. turkey slices
4 oz. potato
3 oz. dressing
2 oz. cranberries

The portions to be served are actually weighed. Some places, such as one big airline catering kitchen, weigh every piece of meat served.

Others spot check; that is, they weigh every twentieth or fiftieth piece served to see that the cook's judgment is good and the proper amount is being put on each plate.

By using a standard recipe, knowing the standard yield to expect from it, and serving the standard portions shown above, the restaurant knows exactly how much it costs to prepare and serve the turkey plate and how much it will have to charge to make a fair profit.

Also, one quick way to make a guest angry is to have him see another guest in the same dining room getting a larger serving than he is for the same price.

Standard Dish-ups

Standard dish-up is a very simple idea, but it is very important. It means putting the food on the plate in the same pattern, or arrangement, each time.

In the case of the turkey plate, the head chef might order the cooks to put the turkey slices on the left side of the plate, the dressing on top of the turkey, the potatoes in the middle, the cranberries on the right side, and a sprig of parsley on the top side.

He knows that this makes the plate look attractive, and he knows that people enjoy a meal much more if it is attractively served with the right combination of placement, quantity, and color on the plate.

Standard Procedures

Standard procedures are followed in all kinds of businesses as well as in well-run hotels, motels, and restaurants. A standard procedure simply means that workers use the same routine for doing a job every time.

Different ways of lining up ingredients for sandwiches, preparing hot food recipes, waiting on tables, making out guest checks, are tried until the best method is found. This method then becomes a standard procedure and all employees do it this way.

All of the standards explained above are designed to eliminate human errors. The customers do not have to depend on the judgment or the mood of different employees each day to get a good bowl of soup or a check properly made out. Standards guarantee the same results every time.

QUESTIONS

1. What do each of the following terms mean?
    a. Standard recipe
    b. Standard yield
    c. Standard portion
    d. Standard dish-up
    e. Standard procedure

2. Why are standards necessary?

CONVERTING STANDARD RECIPES*

A standard recipe produces a standard yield. But—suppose you wish to prepare more or less than the recipe will yield. How do you know how much of the ingredients to use then?

You <u>convert</u> the recipe to yield the amount you desire. For example, if the amount you want to serve is greater than the standard recipe yields, divide the standard yield of the recipe into the amount you want to serve.

Standard yield from recipe = 60
You wish to prepare 300

$$60 \overline{)300} = 5$$

Therefore, the amount you wish to prepare is 5 times greater than the amount the standard recipe yields. Therefore, you would multiply the quantity of each ingredient in the recipe by 5 to get a yield 5 times greater.

Another example: if the amount you want to serve is smaller than the standard recipe yields, then you really only want to serve a fraction of that total amount.

Standard yield from recipe is 6 pans of cake.
You wish to prepare 4 pans of cake.
You want 4/6, reduced to 2/3, of the total recipe yield.

Therefore, you would multiply the quantity of each ingredient in the recipe by 2/3 to get 4 pans of cake instead of 6.

Commercial recipes, those used by restaurants and bakeries, are also called formulas. They are usually written to yield rather large quantities, and they are usually written in measurements of pounds and ounces.

In converting recipes, then, you will be multiplying pounds and ounces by whole numbers or fractions. The easiest way to do this arithmetic is to change any pound amounts to ounces before you start to multiply. Then you will be multiplying only one thing, ounces.

16 ounces = 1 pound

For example: to multiply 1 lb. 8 oz. by 3/4, convert the 1 lb. 8 oz. to 24 oz.

$$\frac{24}{1} + \frac{3}{4} = 18 \text{ oz.}$$

---

* Note that in converting certain recipes, it may not be practical to increase or decrease the quantity of each ingredient by the same rate. For example, in the recipe for Hungarian Goulash, in which you use 4 ounces of garlic for 100 portions, you may not have to use twice as much garlic to make twice as much Goulash. The flavor of garlic is so strong that an additional ounce or two might be sufficient to retain the proper flavor, and doubling the garlic to 8 ounces might ruin the food. Recipes for baked goods cannot always be increased by the same rate either. These decisions can only be reached after actual tests of the receipe at different quantities.

Many recipes can be converted at a constant rate for all ingredients, however; and we can assume that this is true in doing the arithmetic for each of the problems in this lesson.

Then, convert the answer back to pounds and ounces.

18 oz. = 1 lb. 2 oz.

Example of a recipe which produces a standard yield of Graham Cracker Crust for 6 cake pans being converted to a yield of 4 cake pans of crust.

Ingredients for 6 Pans		Amount of Conversion	Amounts Needed to Yield 4 Pans
Crushed graham crackers	2 lbs. (32 oz.)	2/3	21⅓ oz. or 1 lb. 5⅓ oz.
Melted butter	10 oz.	2/3	6⅔ oz.
Granulated sugar	8 oz.	2/3	5⅓ oz.

PROBLEMS

1. The following recipe yields 50 portions of Roast Carre of Lamb. Convert it to yield 150 portions.

Ingredients for 50 Portions		Amount of Conversion	Amounts Needed to Yield 150 Portions
Lamb Racks	50 lbs.		
Salt	2 oz.		
White Pepper	1/2 oz.		
Carrots	1½ lbs.		
Celery	1 stalk		
Onions	2 lbs.		
Brown Stock	1 gal.		
Red Wine	1 pt.		
Cornstarch	2 oz.		
Rosemary	1/2 oz.		

2. The following recipe yields 100 portions of Hungarian Goulash. Convert it to yield 75 portions.

Ingredients for 100 Portions		Amount of Conversion	Amounts Needed to Yield 75 Portions
Boned Chuck, 1-in. squares	35 lbs.		
Oil	1 pt.		
Onions, finely chopped	10 lbs.		
Garlic, finely chopped	4 oz.		
Paprika	12 oz.		
Flour	1½ lbs.		
Salt	4 oz.		
Tomato Puree	2 #10 cans		
White Stock	1 gal.		

3. The following recipe yields 6 tube pans, 4″ × 9½″, of Angel Cake. Convert it to yield 4 pans.

Ingredients for 6 Pans		Amount of Conversion	Amounts Needed to Yield 4 Pans
Egg Whites	4 lbs.		
Sugar	3 lbs. 12 oz.		
Cream of Tartar	1 oz.		
Salt	1/2 oz.		
Vanilla	2 oz.		
Almond Flavor	1/2 oz.		
Cake Flour	1 lb. 8 oz.		

4. The following recipe yields 12 dozen Cheddar Cheese Rolls. Convert it to yield 9 dozen rolls.

Ingredients for 12 Dozen		Amount of Conversion	Amounts Needed to Yield 9 Dozen
Water, 115°	1 lb.		
Dry Yeast	4 oz.		
Sugar	6 oz.		
Salt	2 oz.		
Dry Skim Milk Solids	6 oz.		
Eggs	1 lb.		
Butter, softened	6 oz.		
Bread or All-Purpose Flour	6 lb. 8 oz.		
Cheddar Cheese	8 oz.		

5. The following recipe yields 8 dozen Soft Dinner Rolls. Convert it to yield 5 dozen rolls.

Ingredients for 8 Dozen		Amount of Conversion	Amounts Needed to Yield 5 Dozen
Yeast	4 oz.		
Lukewarm Water	2 lbs. 6 oz.		
Sugar	6 oz.		
Salt	1 oz.		
Nutmeg	pinch		
Egg Solids or Whole Eggs	8 oz.		
Milk Powder	4 oz.		
Butter or Shortening	10 oz.		
Bread Flour	4 lbs.		

AS PURCHASED COMPARED WITH EDIBLE PORTION

Inexperienced cooks are often shocked when they discover what a small amount of food remains after it has been cooked as compared with the amount that was purchased originally. They have to learn that there is almost always a difference between the quantity of food "as purchased" and the yield of "edible portion" after preparation and cooking. These culinary terms mean just what they say: "As purchased" is the weight or volume of food at the time it was purchased, before being cleaned or trimmed in any way. "Yield" means to produce or give. "Edible portion" means the amount of food that can be served and eaten.

YIELD OF ROAST TURKEY

A good illustration of the yield of edible portion from the as purchased quantity is seen when a turkey is bought for a family dinner. Many people buy one pound of turkey for each person they plan to serve. They know that each person will not eat a whole pound, but they also know that the edible portion yield on a roast turkey is about 50%. Thus, if they plan to serve 14 people, they will buy a 14 lb. bird and end up with 7 lbs. of edible turkey to serve, or ½ lb. per person. If they stop to think about it, they will also realize that the turkey which cost $1.50 per lb. as purchased actually costs $3.00 per lb. as served to the guests in the form of edible portions.

YIELD VARIATIONS

Yield percentages of foods will vary somewhat for different people doing the preparation. One person might remove more of the as purchased food than another person. Therefore, yield percentages are approximate.

There are two levels of yield to be aware of. One is the yield of food remaining after an as purchased amount has been cleaned and prepared prior to cooking; the other level is the yield after cooking. Carrots are a good example. As purchased, carrots yield an average 70% edible portion of raw carrots after cleaning and trimming. If they are cooked, shrinkage will bring the weight down a little further, to about 60% of the original as purchased quantity.

AS PURCHASED AND EDIBLE PORTION IN STANDARD RECIPES

In standard recipes, such as those on page 136, some of the ingredients are shown in as purchased quantities and prices, and others are shown in edible portion quantities and prices. In the recipe for Hungarian Goulash the oil, paprika, flour, salt, and tomato puree are as purchased. The boned chuck, onions, and garlic are shown in edible portion quantities. These are the quantities called for after each ingredient has been trimmed and cut.

You might have to purchase 40 lbs. of boned chuck to trim and cut into 35 lbs. of 1-inch squares. This would be a yield of approximately 88%. The 40 lbs. would have cost about $1.64 per lb. as purchased, but after the waste in preparation, it results in a cost of $1.88 per lb. for the 35 lbs.

Whether the quantity and cost of each ingredient are as purchased or as edible portion going into the recipe does not matter. You have the true quantity and cost of each ingredient as it is used in this particular recipe.

By accurately recording the prices as explained above, you will have the correct extension (cost) for each ingredient; and you will get an accurate total cost and cost per portion for the food ready to serve.

HOW TO FIND YIELD PERCENTAGES

Average yield percentages of a long list of foods can be found in guidebooks for professional cooking. Yield percentages can also be established from experience. To find the yield percentage of food you have prepared, divide the edible portion quantity by the as purchased quantity as follows:

celery: 2 lbs. A.P. (as purchased); 1.48 lbs. E.P. (edible portion)

$$2 \overline{)1.48} = .74 = 74\% \text{ yield}$$

THE VALUE OF YIELD PERCENTAGES

Yield percentages are valuable for a number of reasons:

1. When you know the yield percentage and the quantity of food as purchased, you can figure the quantity of edible portion to expect after preparing and cooking.

2. When you know the yield percentage and the quantity of edible portion you desire, you can figure the quantity of as purchased food you need to start with.

3. When you know the yield percentage and the cost of the food as purchased, you can determine the real cost of the edible portion.

Here Are Some Examples:

1. As purchased, breast of chicken fryers yield 66% cooked edible portion. From 10 lbs. of breasts as purchased, what quantity of edible portion can we expect to get?

   Solution: A.P. × yield = E.P.
   66% = .66

   ```
   10 lbs. A.P.
   ×.66
   ────
     60
    60
   ────
   6.60 lbs. E.P.
   ```

2. As purchased, whole chicken for stewing yields 36% edible portion cooked without skin, neckmeat, and giblets. To serve 20 lbs. edible portion, how much stewing chicken do we have to purchase?

Solution: E.P. ÷ yield = A.P.
36% = .36

```
              55.55 lbs. A.P.
      .36)20.0000
          180
          ---
          200
          180
          ---
          200
          180
          ---
          200
          180
          ---
           20
```

As purchased whole chicken for stewing yields 41% edible portion with neckmeat and giblets included. To serve 20 lbs. edible portion with neckmeat and giblets, how much stewing chicken do we have to purchase?

Solution: E.P. ÷ yield = A.P.
41% = .41

```
              48.78 lbs. A.P.
      .41)20.0000
          164
          ---
          360
          328
          ---
          320
          287
          ---
          330
          328
          ---
            2
```

3. As purchased boneless ham yields 63% edible portion cooked. If boneless ham costs $5.00 per lb. as purchased, what is the cost per lb. of edible portion?

Solution: Cost A.P. ÷ yield = cost E.P.
63% = .63

```
              7.936 = $7.94 per lb. E.P.
      .63)$5.00000
           441
           ---
           590
           567
           ---
           230
           189
           ---
           410
           378
           ---
            32
```

QUESTIONS

1. What does the culinary term "as purchased" mean?
2. What does the culinary term "yield" mean?
3. What does the culinary term "edible portion" mean?
4. Why is the edible portion weight or volume smaller than the as purchased quantity?
5. What is the approximate yield percentage of roast turkey?

6. Are the ingredients listed in recipes usually shown as purchased or edible portion?

7. When you know the yield percentage of a food item and you know the quantity of edible portion you want, how do you determine how much food to purchase?

PROBLEMS

Write the formula for finding the edible portion when you know the as purchased quantity and the yield percentage, then do the following 5 problems.

1. The cooked yield percentage for carrots is 60%. What will be the edible portion amount from 20 lbs. as purchased?

2. The cleaned raw yield percentage for carrots is 70%. What will be the edible portion weight from 22 lbs. as purchased?

3. Cauliflower yields 61% cooked. What will the edible portion be from 15 lbs. as purchased?

4. A boneless roll of raw turkey yields 53% cooked. What will be the edible portion from 16 lbs. as purchased?

5. Pork sausage links yield 47% cooked. What will the edible portion weight be from cooking 8 lbs. as purchased?

Write the formula for determining the as purchased quantity when you know the edible portion amount desired and the yield, then do the following 5 problems.

6. As purchased boneless chuck pot roast yields 60% cooked edible portion. How much as purchased roast is needed to get 10 lbs. of edible portion.

7. As purchased rib lamb chops yield 46% of cooked lean meat edible portion. To get 8 lbs. of edible portion, how much as purchased is needed?

8. As purchased ribeye roast yields 70% lean cooked edible portion. How much as purchased roast is needed to yield 15 lbs. of edible portion.

9. As purchased raw shrimp in the shell yields 54% edible portion. How much as purchased shrimp is needed to yield 12 lbs. edible portion?

10. As purchased corn on the cob with husks yields 33% edible portion cooked. How much as purchased corn is needed to yield 30 lbs. edible portion?

Write the formula for determining the edible portion cost when you know the cost as purchased and the yield, then do the following 5 problems.

11. As purchased pork loin roast with bone yields 41% edible portion. If the loin roast costs $1.50 per lb. as purchased, what is the cost per lb. edible portion?

12. Ground beef with no more than 30% fat yields 70% edible portion. If the ground beef costs $2.00 per lb. as purchased, what is the cost per lb. edible portion?

13. Beef heart costs $1.39 per lb. as purchased. The yield on beef heart is 44% cooked edible portion. What is the cost of beef heart per lb. edible portion?

14. Fresh asparagus costs $2.90 per lb. as purchased. The yield on cooked fresh asparagus is 50% edible portion. What is the cost per lb. edible portion?

15. Baking potatoes cost 24¢ per lb. as purchased. The yield for baked potatoes without skin is 74% edible portion. What is the cost per lb. edible portion?

A meal is to be prepared for 50 people. We have selected a menu and the portion sizes to be served. We will need to know:

A. The total edible portion of each item (Tot. Lbs. 50 E.P.)

B. The total quantity of each item to purchase (Tot. Lbs. A.P.)

C. The total cost of each item (Tot. Cost A.P.)

D. The total cost for all items (Total Cost for 50)

E. The cost per portion.

Copy the information as arranged below and fill in the missing figures.

Food	Size E.P.	Tot. Lbs. 50 E.P.	Yield	Tot. Lbs. A.P.	Unit Cost A.P.	Tot. Cost A.P.
16. Pot Roast	4 oz.		45%		$1.30 lb.	
17. Mash Pot.	4 oz.		74%		.25 lb.	
18. Turnips	3 oz.		78%		.33 lb.	
19. Spinach	2 oz.		80%		.21 lb.	
20. Berries	4 oz.		88%		1.33 lb.	
21.					Total Cost for 50	
22.					Cost per Portion	

COSTING STANDARD RECIPES

When a recipe is being standardized for use in a particular restaurant, the cost of all the ingredients that go into the recipe is totalled up and a per-unit cost is figured out. That is, if the recipe yields (makes) fifty chicken pies, the cost of all the ingredients needed to make the fifty pies is added up to get the cost of the recipe; and that total cost is divided by fifty to find the unit cost, the cost of one portion of chicken pie.

Knowing the exact per-unit cost of a food item helps the food and beverage controller, or manager, decide how much should be charged for it when it is served.

Prices of ingredients will change from time to time, so it is necessary to keep an eye on them and refigure the unit cost as the prices change. If the costs change too much, it might be necessary to change the menu price.

The cost sheets of all food and beverage establishments will not look exactly alike, but they will be similar and will all provide the same information—how much it costs to produce the chicken pie, or turkey sandwich, or plate of spaghetti, or whatever the dish.

Here is what a typical cost sheet for a standard recipe for Ragout of Lamb looks like. (Notice that prices have been rounded off to the tenth of a cent, three places after the decimal.)

Item: Ragout of Lamb
Yield: 13 portions
Size of Portion: 4 oz. meat; 9 oz. total

Standard Recipe No. 25
Cost per Portion: $1.47

Ingredients	*Quantity*	*Market Price*	*Extension (Cost)*
Lamb (¾ in. pieces)	6 lbs.	$2.85 per lb.	$17.10
Roux: butter	8 oz.	2.25 per lb.	1.125
flour	8 oz.	.19 per lb.	.095
Lamb stock (bones)	1 lb. 12 oz.	no charge	—
Carrots (½ in. slices)	12 oz.	.25 per lb.	.188
Celery (½ in. slices)	12 oz.	.32 per lb.	.24
Onions (½ in. slices)	12 oz.	.50 per lb.	.375

Total Cost $19.123
Cost per Portion $1.471 or $1.47

PROBLEMS

Problems on costing standard recipes: The quantities and current prices for the ingredients are given in each of the following standard recipes. Find the per-unit cost as follows (do your figuring on binder paper):

A. Extend the quantity and market price figures on each line; that is, multiply the quantity by the price.

B. Total the extension column to find the total cost of the recipes.

C. Divide the total cost by the number of portions the recipe yields to find the cost per portion.

1. *Item: Roast Carre of Lamb*
   Yield: 50 portions
   Size of Portion: 1 rib chop

   Standard Recipe No. 26
   Cost per Portion:

Ingredients	Quantity	Market Price	Extension (Cost)
Lamb racks	50 lbs.	$4.69 per lb.	
Salt	2 oz.	.015 per oz.	
White pepper	½ oz.	.18 per oz.	
Carrots	1½ lbs.	.28 per lb.	
Celery	10 oz.	.016 per oz.	
Onions	2 lbs.	.20 per lb.	
Brown stock	1 gal.	n/c	
Red wine	1 pt.	6.50 per gal.	
Cornstarch	2 oz.	.03 per oz.	
Rosemary	½ oz.	.44 per oz.	

Total Cost _____
Cost per Portion _____

2. *Item: Hungarian Goulash*
   Yield: 100 portions
   Size of Portion: 8 oz.

   Standard Recipe No. 27
   Cost per Portion:

Ingredients	Quantity	Market Price	Extension (Cost)
Boned chuck 1-in. squares	35 lbs.	$2.69 per lb.	
Oil	1 pt.	.64 per pt.	
Onions, fine chopped	10 lbs.	.30 per lb.	
Garlic, fine chopped	4 oz.	.19 per lb.	
Paprika	12 oz.	.50 per lb.	
Flour	1½ lbs.	.26 per lb.	
Salt	4 oz.	.015 per oz.	
Tomato Puree	2 #10 cans	.84 per can	
White stock	1 gal.	n/c	

Total Cost _____
Cost per Portion _____

3. *Item: Angel Cake*
   Yield: 6 pans, 4-in. by 9½ in., 60 portions
   Size of Portion: 10 per pan

   Standard Recipe No. 35
   Cost per Portion:

Ingredients	Quantity	Market Price	Extension (Cost)
Egg whites	4 lbs.	$ .426 per lb.	
Sugar	3¾ lbs.	.52 per lb.	
Cream of tartar	1 oz.	.29 per oz.	

Ingredients	Quantity	Market Price	Extension (Cost)
Salt	½ oz.	.015 per oz.	
Vanilla	2 oz.	.46 per oz.	
Almond flavor	½ oz.	.29 per oz.	
Cake flour	1½ lbs.	.24 per lb.	

Total Cost _____
Cost per Portion _____

4. *Item: Cheddar Cheese Rolls*
   Yield: 12 doz. rolls, 144 portions
   Size of Portion: 3-in. rolls

   Standard Recipe No. 41
   Cost per Portion:

Ingredients	Quantity	Market Price	Extension (Cost)
Water, 115°	1 lb.	$ n/c	
Dry yeast	4 oz.	.14 per oz.	
Sugar	6 oz.	.23 per lb.	
Salt	2 oz.	.015 per oz.	
Milk solids, skim	6 oz.	.80 per lb.	
Eggs	1 lb.	.84 per lb.	
Butter	6 oz.	2.25 per lb.	
All-purpose flour	6½ lb.	.26 per lb.	
Cheddar cheese	8 oz.	2.29 per lb.	

Total Cost _____
Cost per Portion _____

5. *Item: Soft Dinner Rolls*
   Yield: 8 doz., 96 portions
   Size of Portion: 3-in. roll

   Standard Recipe No. 38
   Cost per Portion:

Ingredients	Quantity	Market Price	Extension (Cost)
Yeast	4 oz.	$ .14 per oz.	
Lukewarm water	2⅜ lbs.	n/c	
Sugar	6 oz.	.23 per lb.	
Salt	1 oz.	.015 per oz.	
Nutmeg	pinch	n/c	
Egg solids	8 oz.	.49 per lb.	
Milk powder	4 oz.	.68 per lb.	
Butter	10 oz.	2.25 per lb.	
Bread flour	4 lbs.	.26 per lb.	

Total Cost _____
Cost per Portion _____

ADDITIONAL PRACTICE MULTIPLYING OUNCES AND POUNDS

Did you have difficulty extending some of the quantities and prices in the recipes above because the quantity was shown in ounces and the price was shown per pound?

One simple way to extend (multiply) an item is to convert one of the figures to the same unit of measure as the other. You can convert the quantity from ounces to pounds. Then you will be multiplying a quantity of pounds times the price per pound, like the other items in the recipes.

To convert ounces to pounds, you *divide the number of ounces* by 16 because there are 16 ounces in 1 pound. In other words, you make a fraction, using the number of ounces as a numerator and 16 as a denominator.

1 oz. = $1/16$ lb.   2 oz. = $2/16$ lb.   3 oz. = $3/16$ lb.
19 oz. = $19/16$ lb.

Here are two examples:

Ingredients	Quantity	Market Price	Extension (Cost)
Egg solids	3 oz.	$ .50 per lb.	?

$$\frac{3}{16} \times \frac{\$.50}{1}$$

$$\frac{3}{\underset{8}{\cancel{16}}} \times \frac{\overset{.25}{\cancel{.50}}}{1} = \frac{.75}{8} = .0937 = \$.094$$

Butter	18 oz.	$2.32 per lb.	?

$$\frac{18}{16} \times \frac{\$2.32}{1}$$

$$\frac{\overset{9}{\cancel{18}}}{\underset{8}{\cancel{16}}} \times \frac{2.32}{1}$$

$$\frac{9}{\underset{1}{\cancel{8}}} \times \frac{\overset{.29}{\cancel{2.32}}}{1} = \$2.61$$

Extend the following quantities and prices:

1. Coffee	6 oz.	$2.53 per lb.	
2. Sugar	12 oz.	.28 per lb.	
3. Milk solids	7 oz.	.84 per lb.	
4. Bread flour	9 oz.	.30 per lb.	
5. Tillamook cheese	4 oz.	3.24 per lb.	
6. Egg solids	12 oz.	.57 per lb.	
7. Egg whites	3 oz.	.426 per lb.	
8. Onions, fine chopped	4 oz.	.36 per lb.	
9. Eggs	8 oz.	.852 per lb.	
10. Boiled ham slices	18 oz.	6.40 per lb.	

COOK'S DAILY REPORTS

To control food, beverage, and money in restaurants, foods workers often fill out daily reports. Each restaurant has its own forms, but they all have the same purpose, to keep a check on the food and beverage prepared and served and the money taken in. It is part of the system to prevent waste and theft.

Daily reports are also valuable in forecasting. "Forecasting" means predicting how much food and beverage will be sold tomorrow, or next week, or next month. This is important, of course, for buying raw foods, hiring employees, etc.

COOK'S PRODUCTION REPORT AND COUNTER REPORT

On the Cook's Production Report illustrated below, the cook has written the number of portions prepared for each food item and the number of portions returned, or left over, at the end of the day. The difference between these two figures is the number sold.

The Counter Report, shown on the next page, is the same as the Cook's Production Report except that the number sold of each item is multiplied by the unit price to find the value sold. The total value shows how much money should have been taken in that day for all the food sold over the counter.

Do you know how to find out how many milkshakes you have sold when they come out of an automatic dispenser and you cannot count how many are left over? The answer is: Start the work shift with a certain number of milkshake cups set out next to the dispenser,

COOK'S PRODUCTION REPORT

DATE: 6-15

Food Item	No. Portions Prepared	No. Returned	No. Sold
Roast Beef	50	17	33
Spaghetti	75	18	57
Halibut	50	3	47
Baked Potatoes	50	8	42
Mixed Vegetables	30	4	26
Tomato Soup	40	16	24

Joe Smith
Name

COUNTER REPORT

DATE: 6-15

Food Item	No. Portions For Sale	No. Not Sold	No. Sold	Unit Price	Value Sold
Hamburgers	100	7	93	$1.25	$116.25
Hot dogs	125	12	113	.90	101.70
Milk	100	17	83	.30	24.90
Milkshakes	150	28	122	.70	85.40
Cokes	200	41	159	.35	55.65
Pies	75	12	63	.50	31.50
Potato Chips	75	14	61	.25	15.25

Total Value $430.65

Name: Sally Jones

for example, 100. At the end of the work shift, count how many cups are remaining. The number of cups missing is the number of milkshakes served.

You can count servings of soup, spaghetti, and other things that come out of large containers in the same way. Simply count the number of dishes or bowls you start with, and then count the number you finish with. The missing dishes tell how many you have served.

PROBLEMS

Draw up some blank forms for Cook's Production Reports and for Counter Reports like the ones shown. Fill in the reports from the information given below.

Cook's Production Reports

1. June 15,—

	Prepared	Returned
Roast lamb	50	3
Roast beef	75	6
Filet of sole	40	4
Mashed potatoes	60	0
Rice	45	3
Mixed vegetables	100	6

2. June 16,—

	Prepared	Returned
Roast turkey	60	4
Hamburger steak	50	12
Irish stew	80	0
Green peas	100	7
Baked potatoes	50	2
Mashed potatoes	60	5
Clam chowder	40	4

3. June 17,—

	Prepared	Returned
Roast chicken	40	6
Club steak	35	4
Hamburger steak	50	0
Peas and carrots	75	7
Spinach	50	3
Tomato soup	30	2

Counter Reports

	For Sale	Not Sold	Unit Price
4. July 1,—			
Hamburgers	75	12	$1.25
Hot dogs	75	7	.90
Milk	100	14	.30
Milkshakes	100	23	.70
Cokes	100	48	.35
Pies	50	12	.50
5. July 2,—			
Hamburgers	175	23	$1.25
Pizzas	135	18	.90
Milkshakes	200	28	.70
Cokes	150	17	.35
Corn chips	84	76	.25
6. July 3,—			
Hamburgers	125	38	$1.25
Hot dogs	96	18	.90
Milk	72	8	.30
Cokes	100	39	.35
Pies	48	11	.50

QUESTIONS

1. What is the main purpose of daily reports?

2. What else are daily reports used for?

3. What is meant by the term "forecasting?"

4. What three things does the Cook's Production Report show when it is completed?

5. What does the Counter Report show that the Cook's Production Report does not show?

6. Suppose you were selling colas from a metal 100-drink container. How could you tell how many colas you sold during the day and how many were left when you could not see inside the container?

REPORTS WITH A DOUBLE CHECK

One important thing to remember in keeping business records is to double check your work before giving it to the manager or another employee. The person who reads and uses your reports expects them to be 100 percent accurate and dependable.

Some reports are designed so that there is an automatic double check on the mathematics used. The cost report for a pastry and bake shop shown on this page is one of that type.

Here is how the report works:

a. The cost of each kind of food supply used is written in the proper column each day.

b. Each line is added across. This shows the total cost of food used for each day.

c. At the end of the month, the Daily Total Cost column is added down. This shows the total cost of all food supplies used for the month, $354.55. (To save space, the illustration below shows figures for ten days instead of a complete month.)

d. The individual food columns are added down. This shows the total cost of each individual food item such as milk products, eggs, used for the month.

e. The report is then double checked by:

Adding across the bottom. The totals of all the individual food columns should equal the total of the Daily Total Cost column in the lower left corner, $354.55.

If the first column on the left and the figures across the bottom do not total the same, there is an error in addition somewhere and it should be corrected before giving the report to the manager or anyone else.

PASTRY AND BAKE SHOP—Food Cost

Date Sept.	Daily Total Cost	Milk Products	Eggs	Butter	Shortening	Sugar	Flour	Fruit	Misc.
1	$52.47	6.00	3.69	1.50	5.72	3.60	6.30	22.63	3.03
2	27.43	5.04	3.48	1.25	4.82	2.70	5.17	3.47	1.50
3	42.17	8.71	4.12	2.28	8.00	4.94	8.95		5.17
4	49.83	7.20	4.23	1.80	6.95	3.82	7.47	18.36	
5	33.06	7.09	4.16	1.75	6.82	3.77	7.14		2.33
6	28.41	6.12	3.71	1.53	5.85	3.14	6.41		1.65
7	34.58	4.91	3.45	1.48	4.73	2.56	5.00	12.45	
8	28.18	6.22	3.82	1.60	6.04	3.15	6.38		.97
9	24.83	5.37	3.37	1.27	5.12	2.73	5.49		1.48
10	33.59	5.59	3.50	1.30	5.16	2.85	5.74	7.32	2.13
Total	354.55	62.25	37.53	15.76	59.21	33.26	64.05	64.23	18.26

PROBLEMS

Complete the reports shown below by adding the columns across and down. Double check your work by adding the totals of the different kinds of food supplies across the bottom. They should equal the total of the Daily Total Cost column. (It is not necessary to write in this book. You can write all of your totals down neatly on a separate piece of paper in the same order as they would appear on the reports below.)

PASTRY AND BAKE SHOP—Food Cost

Date Oct.	Daily Total Cost	Milk Products	Eggs	Butter	Shortening	Sugar	Flour	Fruit	Misc.
1		6.10	3.48	1.39	5.80	3.63	6.32	14.87	2.05
2		7.53	3.87	1.55	6.15	4.02	7.84		3.04
3		8.10	4.07	2.12	7.03	4.38	8.56	10.50	
4		5.92	2.89	1.41	5.66	2.54	6.25	7.12	1.73
5		6.22	3.60	1.41	5.98	3.65	6.44		1.85
6		7.21	4.59	2.40	6.97	4.64	7.43	2.27	2.94
7		7.12	4.46	2.27	6.84	4.52	7.31	16.73	.77
8		6.75	4.24	2.25	6.48	4.33	7.05	11.08	3.78
9		6.04	3.73	1.54	5.76	3.64	6.34		5.81
10		5.95	3.64	1.45	5.07	3.55	6.25	12.10	3.00
Total									

PASTRY AND BAKE SHOP—Food Cost

Date Nov.	Daily Total Cost	Milk Products	Eggs	Butter	Shortening	Sugar	Flour	Fruit	Misc.
1		6.08	3.46	1.37	5.78	3.61	6.30	14.08	2.03
2		6.15	3.53	1.44	5.85	3.68	6.29	6.07	1.98
3		7.00	4.38	2.29	6.70	4.53	7.22		2.52
4		6.58	3.96	1.87	6.28	4.11	6.80	9.20	2.53
5		5.75	3.21	1.12	5.53	3.36	6.05	19.17	.82
6		6.04	3.42	1.33	5.74	3.57	6.26	11.10	1.58
7		6.20	3.56	1.47	5.88	3.01	6.47		4.15
8		6.05	3.43	1.34	5.74	3.58	6.27	12.63	1.06
9		6.50	3.86	.98	6.92	4.00	6.80	13.03	.67
10		8.01	5.45	3.38	7.77	5.60	8.31	7.58	2.24
Total									

ACCUMULATED EXPENSE AND SALES REPORTS

Pictured on this page are four more examples of daily reports. Figures from the Snack Bar report and the Fountain report are combined to make the Summary of All Departments to Date report. Figures from the Summary of All Departments to Date report plus figures from the reports on the same day last month and last year are combined to make the Comparative Summary.

The Comparative Summary is important for keeping an eye on the trend of the business: Is business better or worse today than it was at this time last month or last year?

ACCUMULATED EXPENSES AND SALES—SNACK BAR

	Cost of Sales		Payroll		Other Exp.		Total Exp.		Total Sales		Net Profit	
Brought Forward	750	00	625	25	24	50	1399	75	1550	00	150	25
Today	245	50	230	50	18	00	494	00	560	75	66	75
To Date	995	50	855	75	42	50	1893	75	2110	75	217	00
% of Sales Today	43	8	41	1	3	2	88	1	100	0	11	9
% of Sales from 1st	47	2	40	5	2	0	89	7	100	0	10	3

ACCUMULATED EXPENSES AND SALES—FOUNTAIN

	Cost of Sales				Payroll		Other Exp.		Total Exp.		Total Sales		Net Profit	
	Fnt. Prep.		Pkg. Gds.											
Brought Forward	800	00	680	50	1220	25	32	00	2732	75	3150	00	417	25
Today	224	35	210	15	405	20	8	50	848	20	969	25	121	05
To Date	1024	35	890	65	1625	45	40	50	3580	95	4119	25	538	30
% of Sales Today	23	1	21	7	41	8		9	87	5	100	0	12	5
% of Sales from 1st	24	9	21	6	39	5	1	0	86	9	100	00	13	1

SUMMARY OF ALL DEPARTMENTS TO DATE

Cost of Sales		%	Payroll Exp.		%	Other Exp.		%	Total Exp.		%	Sales		Net Profit		%
2910	50	46.7	2481	20	39.8	83	00	1.3	5474	70	87.9	6230	00	755	30	12.1

COMPARATIVE SUMMARY

	To Date		Last Month		Last Year	
Total Sales	6230	00	6155	10	6082	35
Total Expenses	5474	70	5410	05	5394	60
Profit	755	30	745	05	687	75
Profit %	12	1	12	1	11	3

PROBLEMS

Below are three reports like those shown on the previous page. The Fountain report needs to be completed. Then, figures from the Snack Bar report and the Fountain report need to be combined to complete the Summary of All Departments to Date report.

Make column headings like those below on a piece of blank paper and write the figures needed to complete these reports. Proceed as follows:

Accumulated Expenses and Sales—Fountain report:

1. To get the answers for the To Date line, add the Brought Forward line and the Today line.

2. To get the % of Sales Today answers, divide the Total Sales Today into each of the other figures on the Today line. (Figure percentages to tenths of 1 percent.)

3. To get the % of Sales from 1st answers, divide the Total Sales to Date into each of the other figures on the To Date line.

Summary of All Departments to Date report:

1. To get the figures for the Summary, add the figures from the Snack Bar and the Fountain reports from the same columns.

2. To get the percentages for the Summary, divide the Sales figure in the Summary into each of the other figures in the Summary.

ACCUMULATED EXPENSES AND SALES—SNACK BAR

	Cost of Sales		Payroll		Other Exp.		Total Exp.		Total Sales		Net Profit	
Brought Forward	850	00	725	25	26	50	1601	75	1772	80	171	05
Today	257	80	242	60	27	50	527	90	694	85	166	95
To Date	1107	80	967	85	54	00	2129	65	2467	65	338	00
% of Sales Today	37	1	34	9	4	0	76	0	100	00	24	0
% of Sales from 1st	44	9	39	2	2	2	86	3	100	00	13	7

ACCUMULATED EXPENSES AND SALES—FOUNTAIN

	Cost of Sales				Payroll		Other Exp.		Total Exp.		Total Sales		Net Profit	
	Fnt. Prep.		Pkg. Gds.											
Brought Forward	928	35	807	50	1350	70	32	80	3119	35	3508	50	389	15
Today	285	80	256	15	429	85	29	75	1001	55	1129	40	127	85
To Date														
% of Sales Today														
% of Sales from 1st														

SUMMARY OF ALL DEPARTMENTS TO DATE

Cost of Sales		%	Payroll Exp.		%	Other Exp.		%	Total Exp.		%	Sales		Net Profit		%

153

BREAK-EVEN ANALYSIS (TO FIND BREAK-EVEN POINT IN SALES)

A break-even analysis is a method used by restaurant owners to determine how much in sales is needed to offset all their expenses and at least break even financially. To do this analysis, they must know what their fixed expenses are, and they must be able to estimate fairly accurately what their variable expenses will be.

Many restaurants that have gone broke after only a few months of operation would never have been started in the first place if the owners had done a break-even analysis and discovered that the costs were so high there was no chance of bringing in enough sales to cover the costs, or break even.

FIXED EXPENSES

Fixed expenses in restaurants are things such as rent, insurance, depreciation on equipment, interest on loans, and property taxes. These expenses are the same every month, no matter how many days a month the restaurant is open or how many meals are served. The expenses are fixed and the owners know how much they will be before the month starts.

VARIABLE EXPENSES

Variable expenses are those costs that vary, or change, as the volume of business varies. The more food you sell, the more food you have to buy and prepare. More sales require more labor, more supplies, more gas and electricity, more laundry, and more of all the other costs that go with preparing each meal.

From experience, restaurant owners learn approximately what their variable expenses will be in terms of percentages of sales. For example, they might find that, when they are operating efficiently, the cost of food and beverage purchases runs about 34% of sales. That means that it costs them 34¢ to buy the food and beverage that they sell for one dollar, or a cost of $34 for $100 of sales. They also learn what percent of each sales dollar they must spend for labor, laundry, utilities, cleaning, and all the other routine variable expenses. Adding up the percentages for these variable costs, the owners can come up with a total percentage that gives them a fairly good idea of what it costs to prepare and serve various volumes of sales.

Different types and sizes of restaurants might have different percentage figures for their total variable costs. One might have 62%; and another, 68%. Once the manager can calculate what the fixed costs are and what the percentage of variable costs usually is, he or she can determine a break-even point in sales: that is, the manager knows how much money the restaurant has to take in to pay all the expenses, with no profit or loss. This is called making a break-even analysis, or figuring a break-even point.

TIME PERIOD FOR ANALYSIS

The break-even analysis can be calculated for different periods of time, such as a day, a week, a month, or longer. A month is a good period to use because it fits in with other accounting practices and gives a reasonable time for slow and busy days to even out.

NEW BUSINESSES

It is easier, of course, to make a break-even analysis for a business that has been operating for a while than for a new business that has not yet opened, because you have a history of figures on which to base your fixed and variable costs. Even for a new business, however, where some estimates have to be made without knowing exactly what the figures will be, the break-even analysis can be very helpful to the investors in predicting success or failure.

THE ANALYSIS—HOW COMPLICATED?

Doing a break-even analysis can become quite complicated and technical; and big businesses, such as restaurant chains, have the knowledge and equipment to do this. Small restaurant operators can do a rather simple analysis which might not be as accurate as the more complicated one, but nevertheless can be very useful.

THE BASIC FORMULA

The basic idea or formula for determining how much in sales is needed to break even when the fixed costs are known and the variable costs percentage can be estimated is as follows:

1. The fixed expenses for a month are added up. Let us say they total $12,000.

2. The variable expenses percentage is calculated. Let us say that past figures show that the total variable expenses average 68% of sales.

3. The percentage of variable expenses, 68%, is subtracted from 100% (which is total sales); that leaves 32% to pay for the fixed costs.

4. The fixed costs are $12,000, and they are 32% of sales (100%). So the remaining question is: $12,000 is 32% of what? That is the amount of total sales needed to break even. To find what $12,000 is 32% of, divide the $12,000 by 32% as follows:

```
                          $   37500.
    32% = .32          .32)$12,000.00
                           96
                           240
                           224
                            160
                            160
                            000
```

This arithmetic shows that, with fixed costs of $12,000 and variable costs of 68%, you need sales of $37,500 to break even. You can prove your answer as follows:

Multiply	Multiply
$37,500 Sales	$37,500 Sales
.32 Fixed cost percentage	.68 Variable cost percentage
75000	300000
112500	225000
$12,000.00	$25,500.00

Add

$12,000.00 Fixed costs
25,500.00 Variable costs
$37,500.00 Total costs on sales of $37,500 = break-even point

QUESTIONS

1. What is a break-even analysis?
2. What are fixed expenses?
3. What are variable expenses?
4. How do restaurant owners learn how much their variable expense percentages will be?
5. Will the variable expenses be the same for all restaurants?
6. For what period of time can a break-even analysis be made?
7. Can you do a break-even analysis for a new restaurant before it opens?

PROBLEMS

Find the break-even point in sales from each of the following sets of figures:

1. Fixed costs $13,600   Variable costs 66% of sales
2. Fixed costs $19,600   Variable costs 65% of sales
3. Fixed costs $16,064   Variable costs 68% of sales
4. Fixed costs $18,315   Variable costs 63% of sales
5. Fixed costs $18,635   Variable costs 64% of sales
6. Fixed costs $12,610   Variable costs 81% of sales
7. Fixed costs $19,127   Variable costs 70% of sales
8. Fixed costs $17,830   Variable costs 71.5% of sales
9. Fixed costs $5,315   Variable costs 73.5% of sales
10. Fixed costs $6,488   Variable costs 79.5% of sales

SEATING TURNOVER

Restaurant operators keep a close watch on the number of covers served at each meal and the number of turnovers. The word "cover" means customer served, also called "guest" in the hospitality industry. The "number of turnovers" means the number of times the total number of seats in the dining room are used. If you had 100 seats and you served 300 covers at lunch, you would have a turnover of 3; that is, you used all 100 seats 3 times for that meal.

PLANNING A NEW RESTAURANT

Experienced people who are thinking of opening a new restaurant do some careful research and planning first. In connection with their break-even analysis, they set a target of a certain number of covers and a certain turnover of seats. These are mathematical tools that help the investors decide whether or not it would be profitable to open a restaurant in a particular location.

THE ENTREPRENEURS' RISK

People who invest their money and operate a new business are known as entrepreneurs. Entrepreneurs in the restaurant business know that there is a considerable risk of losing their money. A high percentage of restaurants show a loss and go out of business within a few months or a year of opening.

Yet, some new restaurants are very successful and profitable. These successful operations are organized by entrepreneurs who have done their homework before spending any money to start up. They study things like the location they are considering for their restaurant. They look at the traffic patterns and the parking situation. They observe the type of people who are in the area, business or residential, what kinds of food their potential guests might like, how much they might spend for a meal, and how often they might be inclined to eat in a restaurant.

The entrepreneurs must also consider whether or not they would have any difficulty finding skilled employees in that area and whether or not there would be any problem getting raw food and supplies delivered to that location. Along with these and many other things to be considered, the entrepreneurs do their arithmetic, which includes their break-even analysis and the estimate of the number of covers to be served at each meal and the turnover of seats.

In a business district, where people are likely to come in soon after work, eat, and go home, one turnover for dinner might be expected. In a residential district, people might arrive for dinner over a more extended period of time, say from 6 P.M. to 9 P.M. In this latter case, 1½ to 2 turnovers might be a reasonable expectation.

In an effort to get some people to come early and make it easier to serve two turnovers, some restaurants offer an "early-bird special." This is a less expensive meal served to guests arriving before 6:30 or 7:00 P.M. It may be limited to 3 or 4 entrees, making it quicker to serve. In addition to spreading out the arrival time of guests, that is, getting some to come earlier, it brings in people who could not afford the regular menu.

RELATING COVERS AND TURNOVERS TO THE BREAK-EVEN ANALYSIS

The break-even analysis can be broken down to a daily amount of sales needed, and the covers and turnovers are related to those daily figures. If you calculate, for example, that you need sales of $1,200 a day, you have to figure out some combination of covers and average check per cover to equal or exceed $1,200. If you have 100 seats and serve dinner only, at one turnover and an average check of $12, you would take in the $1,200 you need to break even. All sales above the $1,200 figure would result in a percentage of profit for the owners.

Let us say that your break-even point was $1,400 a day and you still had 100 seats and served dinner only. At an average of $12 per check, 100 covers would leave you losing $200 a day. Therefore, you would have to figure out some way to sell more covers or increase your average check. This involves things like a little more salesmanship by the employees, a better public relations or advertising campaign, or, perhaps, a change in menu prices.

SALESMANSHIP

Average checks can be increased by a simple thing like selling more desserts and beverages. For example, if you could sell a $4 dessert to every guest, you would increase your average check by $4. If you could sell a $4 dessert to only half of the guests, you would increase your average check by $2. A $3 glass of wine sold to 50% of your guests would increase your average check by $1.50. Successful entrepreneurs do their arithmetic and solve these problems.

QUESTIONS

1. In the restaurant business, what is a cover?

2. What is meant by the number of turnovers?

3. What is an entreprenuer?

4. Is it easy to start a new restaurant and make a profit?

5. List 4 of the things you should consider, or look at, before opening a new restaurant.

6. Name 2 ways owners try to increase their average check.

PROBLEMS

Below are a few combinations of figures for you to work on to get the feel of the relationship between turnovers, covers, average check, and break-even sales. It is very simple. The basic formulas are:

No. of seats × turnovers = covers   Covers × average check = sales

1. 100 seats, 2 turnovers, $13 average check: a.) How many covers served? b.) How much in sales?

2. 120 seats, 1½ turnovers, average check $14: a.) How many covers? b.) How much in sales?

3. 100 seats, 3 turnovers, $3,300 sales needed: How much average check is needed?

4. 60 seats, 2½ turnovers, $1,650 in sales needed: a.) How many covers sold? b.) How much average check needed?

5. 120 seats, 2 turnovers, $3,360 in sales needed: a.) How many covers sold? b.) How much average check needed?

6. 60 seats, 1½ turnovers, $1,890 in sales needed: a.) How many covers sold? b.) How much average check needed?

7. 76 seats, 3 turnovers, $2,166 in sales needed: How much average check needed?

8. $2,700 in sales needed, $18 average check: How many covers needed?

9. $1,125 in sales needed, 60 seats, 1½ turnovers: a.) How many covers served? b.) How much average check needed?

10. $2,480 in sales needed, $16 average check, 2 turnovers: How many seats needed?

11. $2,436 in sales needed, $14.50 average check: How many covers needed?

12. $1,440 in sales needed, average check $8, 3 turnovers: a.) How many covers needed? b.) How many seats needed?

13. $3,840 in sales needed, $24 average check, 2 turnovers: a.) How many covers needed? b.) How many seats needed?

14. 60 seats, 2 turnovers, average check $24: Produces how much in sales?

15. 60 seats, 2 turnovers, $3,000 in sales needed: a.) How many covers served? b.) How much in average check needed?

PRICING THE MENU (MARK-UP)

Who decides on the prices that are shown on the menu? And, how do they decide how much to charge?

In a large organization, such as a company that owns a chain of restaurants or a large hotel, a number of people might help decide what menu prices should be. The final decision would then be made by the Food and Beverage Controller. He/she is the person in charge of all the food and beverage operations and supervises the chefs, maitre d's*, waiters, bartenders, and cashiers. The Food and Beverage Controller is really the chief accountant and is responsible for the profit or loss of the restaurant.

In a giant food and beverage organization that operates many drive-ins, lunch counters, etc., the cost of everything that is purchased, plus the cost of labor, rent, equipment, supplies, taxes, and all the other business expenses, is measured very carefully and then fed into a computer. The computer then tells the executives of the company how much they will have to charge to make a profit.

In a smaller organization, such as an individual restaurant, the manager (who might be the owner), with advice from the chef and the head waiter, would decide how much to charge for various meals. He/she tries to set prices low enough to compete with other eating places and yet has to be careful not to fall into the trap of setting them so low that despite many customers the restaurant ends up losing money anyway.

"Mark-up" is a term used by restaurant and other business people. Mark-up is the cost of doing business expressed by a percentage. An amount equal to this percentage must be added to raw food costs to get a selling price that will cover all the costs of operating a restaurant or hotel diningroom.

For example, a cafeteria might mark up all its meat dishes by one-half the cost to get the selling price. Another restaurant might mark up all meat dishes by two-thirds, while another, fancier restaurant might add a mark-up of two or three times the cost of the raw food to get the menu price.

Here is an example of how the selling price is figured if the raw food cost is 85¢ and the mark-up rate is 2/3.

a) Multiply the food cost by the mark-up rate: 85¢ × 2/3

$$\frac{85}{1} \times \frac{2}{3} = \frac{170}{3} = 56\frac{2}{3}¢$$

b) Add the mark-up to the food cost:

    85¢
  +56⅔
  141⅔¢   or $1.41⅔   or   rounded off, $1.42

The answer, or selling price, is $1.42.

---

* *Maitre d'* is the short form used by people to refer to the Maitre d'hotel, the person in charge of the dining room, waiters, busboys, etc.

PROBLEMS

I. Problems on figuring the amount of mark-up and selling price:

	Raw Food Cost	Rate of Mark-Up	Amount of Mark-Up	Selling Price
1.	$ .96	2/3		
2.	1.44	2/3		
3.	.75	3/4		
4.	.52	1/4		
5.	1.25	1/2		
6.	.75	3/5		
7.	.89	3/5		
8.	1.15	4/5		
9.	.96	5/6		
10.	.47	7/8		
11.	.64	5/8		
12.	.87	5/8		
13.	1.27	1/3		
14.	2.04	2/3		
15.	1.68	3/4		
16.	.70	4/5		
17.	.75	1/2		
18.	.68	3/4		
19.	.35	3/5		
20.	.85	3/8		
21.	.90	1½		
22.	3.70	1½		
23.	4.50	1¼		
24.	1.50	1⅓		
25.	.60	2½		

Mark-up can be figured using percentages, or decimals, as well as by using fractions.

Remember: to use percentage figures in multiplication or division, it is easier if you convert the percentage figure to its decimal equivalent. To change a percent figure to a decimal, you remove the percent sign and move the decimal point two places to the left. Examples:

85% = .85   27.5% = .275   9% = .09

II. Problems on figuring the amount of mark-up and selling price using percentage rates.

	Raw Food Cost	Rate of Mark-Up	Amount of Mark-Up	Selling Price
1.	$ .92	60%		
2.	1.34	65%		
3.	.72	75%		
4.	.48	100%		
5.	1.35	70%		
6.	.83	67%		
7.	1.30	45%		
8.	.98	34%		
9.	.49	33%		
10.	2.05	150%		
11.	1.86	28%		
12.	.78	33.3%		
13.	1.09	42.5%		
14.	.38	27.5%		
15.	1.25	66.6%		
16.	.92	68%		
17.	3.12	55%		
18.	1.27	87.5%		
19.	.62	62.5%		
20.	.85	16.6%		

QUESTIONS

1. Who decides the prices that are shown on the menu in a large food organization?
2. Who decides menu prices in an individual restaurant?
3. What does the term "mark-up" mean?

WHAT IS AN OUNCE?

The word "ounce" has two meanings: as a measure of weight or a measure of volume. Commercial formulas, or recipes, show the weights of the various ingredients to be used in preparing a food item. Commercial cooking is called quantity cooking; that is, food is prepared in large quantities. Weighing the ingredients is considered to be the most accurate way of preparing food in large quantities. Thus, a commercial recipe calling for 8 ounces of sugar means that 8 ounces of sugar should be weighed out on a scale.

Recipes designed for the family kitchen, homemaking recipes, indicate the volume of ingredients needed. They usually show the number of tablespoons or cups to be used. Therefore, a homemaking recipe calling for 1 cup of sugar means that you should fill a measuring cup with sugar up to the 8-ounce mark. The space in 1 cup used in a home recipe is different than the 8 ounces of weight called for in the commercial recipe.

PURCHASING

Most of the items on the shelves of grocery stores show the weight of the package or can, for example an 11-ounce can of soup. There are some items, however, usually liquids, that are sold in terms of volume, also called "fluid ounces." Milk is sold in volume, or fluid ounces, and so are wine and other beverages. When you buy a pint of milk, you are getting 16 fluid ounces of milk, which is the same as 16 volume ounces. On the other hand, when you buy a pound of butter, you are getting 16 ounces weight of butter.

In buying food and beverages and preparing meals, it is important to have a clear understanding of whether you are dealing in weight or volume, especially if the word "ounce" is used.

Following is a summary of abbreviations and equivalents that will help you keep weight and volume straight in your mind.

Abbreviations	*Volume*	*Weight*
tsp. or t. = teaspoon	3 t. = 1T.	16 oz. = 1 lb.
tbsp. or T. = tablespoon	2 T. = 1 oz.	100 lb. = 1 cwt.
c. = cup	8 oz. = 1 c.	2,000 lb. = 1 ton
lb. = pound	2 c. = 1 pt.	
pt. = pint	4 c. = 1 qt.	
qt. = quart	2 pt. = 1 qt.	
oz. = ounce	4 qt. = 1 gal.	
min. = minute	16 oz. = 1 pt.	
hr. = hour	32 oz. = 1 qt.	
f.g. = few grains		
cwt. = hundred weight		
gal. = gallon		

WINE WITH YOUR MEAL?

A *gourmet* is a person who is experienced in judging fine foods and wines. A *gourmand* is a person who may not be an expert judge of food, like a gourmet, but loves to eat and usually eats large quantities. Most gourmets would not think of eating a fine meal without wine, while most gourmands would certainly drink wine with lunch or dinner also.

VARIETIES OF WINE

Wine is made from grapes; and as there are many varieties of grapes grown all over the world, there are many varieties of wine. Many wines get their names from the area where the grapes were first grown.

Champagne is a sparkling wine that is popular for celebrations of weddings, anniversaries, birthdays, or any moment of success at any time of day or night.

The most popular varieties of wines that are enjoyed before dinner are sherry and vermouth. Wines drunk with meals fall into three general categories: white, pink, and red. Some of the most popular white wines with dinner are chablis, sauterne, Rhine, Riesling, and chenin blanc. The most popular dinner red wines include burgundy, cabernet sauvignon, gamay beaujolais, and pinot noir. The most common pink dinner wine is known as vin rose.

THE PLEASURES OF WINE

A fine wine brings out the flavor of good food and makes it taste even better. Some people also feel that wine aids digestion. The variety of wine you select depends on your taste and, perhaps, on your pocketbook. Most people drink a white wine with light foods such as seafood or fowl (chicken, turkey, pheasant, etc.). With the heavier meats, such as steak or roast beef, people usually prefer the heavier red wines. The rose variety, which is in between the heavy red and the light white wines, goes well with most any kind of food.

For after dinner, or at any other time when food is not being served, many people like the sweet-tasting wines such as sherry, port, and muscatel.

WINE LIST OR HOUSE WINE?

When you wish to order wine in a first-class restaurant, the waiter or maitre d' hands you a wine list. This list shows the brands, called labels, and the prices of the various wines that are available for you to buy with your meal. If you are not interested in ordering a particular label, you can usually order the "house wine," which is a good wine that the restaurant buys in large quantities at a discount and can sell to you at a lower price than you would have to pay for one of the well-known labels from the wine list.

BOTTLE SIZES

For years the most popular size for a bottle of wine in the United States has been 4/5 quart, which is 1/5 of a gallon, known simply as "a fifth." A fifth contains 25.6 fluid ounces, calculated by dividing the 128 ounces in a gallon by 5.

(As the United States converts to the metric system, an increasing number of bottles are appearing in such quantities as 750 milliliters, which is almost the same size as a fifth, and 1 liter, which is slightly larger than a quart.)

GLASS SIZES

If you do not want a whole bottle of wine with your meal, you may order a glass. There are various shapes and sizes of wine glasses for different kinds of wines and different occasions. With a meal in a restaurant, your wine would usually be served in a 6-ounce or an 8-ounce glass, or occasionally a 10-ounce glass. There are also wine glasses that hold 20 ounces or more.

For a little fun, get four or five wine glasses of different shapes and sizes and see if you can guess how many ounces each glass holds. You can measure the size by filling each one with water and pouring the water from the glass into a measuring cup.

Remember, when we talk about ounces here, we are talking about liquid volume, or capacity, not ounces of weight. Therefore, it does not matter whether you have wine, or water, or any other liquid in the glass. You are measuring the ounces of volume of the glass, or how much space is in it.

QUESTIONS

1. What is a gourmet?

2. What is a gourmand?

3. What varieties of wine are often drunk before dinner?

4. Are the following dinner wines red, white, or pink?
   a. chablis
   b. burgundy
   c. sauterne
   d. gamay beaujolais
   e. vin rose

5. What kind of wine do most people prefer to drink with
   a. seafood?
   b. fowl?
   c. meat such as steak or roast beef?

6. Name two varieties of wine that are often popular for drinking after dinner.

7. Which is less expensive, wine selected from a wine list or the house wine?

8. a. How many ounces are there in a gallon?
   b. How many ounces are there in a fifth?

PROBLEMS

In the problems below, figure out how many glasses of wine you would get from a bottle by dividing the number of ounces in the glass into the number of ounces in the bottle. Use decimals in dividing.

Show two answers for each problem. In the first answer, stop at two places after the decimal. For the second answer, round the first answer off to a whole number.

How many glasses can be poured from a fifth if the glass size is:

1. 6 ounces?

2. 8 ounces?

3. 10 ounces?

How many glasses can be poured from a gallon if the glass size is:

4. 6 ounces?

5. 8 ounces?

6. 10 ounces?

DRILL FOR BAKER'S SCALE

EQUIPMENT NEEDED:

1. Baker's scale
2. Pan for scale
3. Set of weights
   a. Weight to balance pan
   b. Individual balancing weights:
      1 @ 1 lb.
      1 @ 2 lb.
      1 @ 4 lb.
4. Sand (clean and dry) to represent ingredients to be weighed.
5. Pitcher to put weighed sand into.

Add the ingredients in each formula and write the total below each formula. Weigh the pitcher. Add the weight of the pitcher to the total of each formula.

Using sand to represent each ingredient, weigh out the ingredients one by one in the pan and place them in the pitcher after each one is weighed.

Weigh the pitcher full of ingredients. It should weigh within one ounce of the total of the ingredients and the pitcher as shown on the formula sheet below.

How to Add Pounds and Ounces

There are 16 ounces in a pound. For each 16 ounces in the ounce column, carry 1 lb. to the pound column. For example, in 36 ounces there are 2 lb. and 4 oz. remaining (36 ÷ 16 = 2 r4).

```
              2 lb.  9 oz.
              1 lb. 14 oz.
              4 lb. 13 oz.
              ─────────────
              7 lb. 36 oz.  (CARRY 2 LB., REMAINDER 4 OZ.)
(CARRIED)     2     4
              ─────────────
   ANSWER     9 lb.  4 oz.
```

167

PROBLEMS

Ingredients
and pitcher
actual weight

Formula No. 1

Flour 2 lb. 8 oz.
Shortening 1 lb. 2 oz.
Sugar 1 lb. 6 oz.
Salt ½ oz.

Ingredients
and pitcher
actual weight

Formula No. 2

Flour 2 lb. 9 oz.
Shortening 8 oz.
Sugar 1 lb. 8 oz.
Baking Powder 4 oz.
Salt ¾ oz.

Formula No. 3

Flour 3 lb. 7 oz.
Sugar 1 lb. 10 oz.
Shortening 2 lb.
Baking Powder 9 oz.
Salt 3 oz.
Baking Soda 1¼ oz.

Formula No. 4

Flour 2 lb. 9 oz.
Shortening 1 lb. 8 oz.
Sugar 1 lb.
Baking Powder 4 oz.
Salt 2¼ oz.
Baking Soda ½ oz.

Formula No. 5

Flour 1 lb. 12 oz.
Shortening 5 oz.
Sugar 7 oz.
Salt ¼ oz.

Formula No. 6

Flour 2 lb. 12 oz.
Shortening 1 lb. 10 oz.
Sugar 12 oz.
Salt 1½ oz.
Baking Powder 3 oz.
Baking Soda ¾ oz.

GUEST CHECKS

PREPRINTED

The bill that is given to guests for food or beverage served is called a "meal check" or a "guest check," or sometimes just a "check." The checks are numbered consecutively so that the manager can determine at the end of the day whether each check was paid or whether there were some "skips" or "walk outs" (people who walk out without paying). It is the waiter's responsibility to see that each party served gets a check and pays. Some restaurants even fine their waiters for each missing check. Some unions object to this practice, however. Besides the check number, there is a space where the waiter fills in the date, the number of people served, and his or her own initials, information which is important to the accounting system.

Restaurant Guest Checks

In restaurants and dining rooms where the menu is changed every day, the waiter writes on the blank guest check each item served, the prices, the tax, and the total.

Preprinted Guest Checks

Drive-ins, snack bars, and other restaurants that feature fast service and the same menu every day often use a preprinted form of guest check. These checks have printed on them the names and prices of the food and beverage items on the daily menu. The carhop or attendant simply circles or checks with a pencil the items served and their prices.

Partly Preprinted Checks

Some places use guest checks that have the food and beverage preprinted but not the prices. In these places the attendant writes in the quantities served and the prices.

Preprinted or partly preprinted checks save time and reduce errors, especially when the employees are inexperienced.

How to Complete the Guest Check. The important things to remember when filling in guest checks are:

a. To enter on the check everything that is served, including drinks.

b. To charge the proper prices.

c. To extend the prices correctly if two or more orders of the same item are served.

d. To total the check correctly and add the correct tax.

Sales Tax

The tax is usually determined by referring to a tax table, like the one shown.

Also shown are two copies of a guest check of the partly preprinted kind. One of them shows how the check looks before the attendant completes it, and the other shows how it looks after being properly filled in.

6% TAX SCHEDULE

TO	TAX	TO	TAX	TO	TAX	TO	TAX	TO	TAX
.10 -	.00	5.08 -	.30	10.08 -	.60	15.08 -	.90	20.08 -	1.20
.22 -	.01	5.24 -	.31	10.24 -	.61	15.24 -	.91	20.24 -	1.21
.39 -	.02	5.41 -	.32	10.41 -	.62	15.41 -	.92	20.41 -	1.22
.56 -	.03	5.58 -	.33	10.58 -	.63	15.58 -	.93	20.58 -	1.23
.73 -	.04	5.74 -	.34	10.74 -	.64	15.74 -	.94	20.74 -	1.24
.90 -	.05	5.91 -	.35	10.91 -	.65	15.91 -	.95	20.91 -	1.25
1.08 -	.06	6.08 -	.36	11.08 -	.66	16.08 -	.96	21.08 -	1.26
1.24 -	.07	6.24 -	.37	11.24 -	.67	16.24 -	.97	21.24 -	1.27
1.41 -	.08	6.41 -	.38	11.41 -	.68	16.41 -	.98	21.41 -	1.28
1.58 -	.09	6.58 -	.39	11.58 -	.69	16.58 -	.99	21.58 -	1.29
1.74 -	.10	6.74 -	.40	11.74 -	.70	16.74 -	1.00	21.74 -	1.30
1.91 -	.11	6.91 -	.41	11.91 -	.71	16.91 -	1.01	21.91 -	1.31
2.08 -	.12	7.08 -	.42	12.08 -	.72	17.08 -	1.02	22.08 -	1.32
2.24 -	.13	7.24 -	.43	12.24 -	.73	17.24 -	1.03	22.24 -	1.33
2.41 -	.14	7.41 -	.44	12.41 -	.74	17.41 -	1.04	22.41 -	1.34
2.58 -	.15	7.58 -	.45	12.58 -	.75	17.58 -	1.05	22.58 -	1.35
2.74 -	.16	7.74 -	.46	12.74 -	.76	17.74 -	1.06	22.74 -	1.36
2.91 -	.17	7.91 -	.47	12.91 -	.77	17.91 -	1.07	22.91 -	1.37
3.08 -	.18	8.08 -	.48	13.08 -	.78	18.08 -	1.08	23.08 -	1.38
3.24 -	.19	8.24 -	.49	13.24 -	.79	18.24 -	1.09	23.24 -	1.39
3.41 -	.20	8.41 -	.50	13.41 -	.80	18.41 -	1.10	23.41 -	1.40
3.58 -	.21	8.58 -	.51	13.58 -	.81	18.58 -	1.11	23.58 -	1.41
3.74 -	.22	8.74 -	.52	13.74 -	.82	18.74 -	1.12	23.74 -	1.42
3.91 -	.23	8.91 -	.53	13.91 -	.83	18.91 -	1.13	23.91 -	1.43
4.08 -	.24	9.08 -	.54	14.08 -	.84	19.08 -	1.14	24.08 -	1.44
4.24 -	.25	9.24 -	.55	14.24 -	.85	19.24 -	1.15	24.24 -	1.45
4.41 -	.26	9.41 -	.56	14.41 -	.86	19.41 -	1.16	24.41 -	1.46
4.58 -	.27	9.58 -	.57	14.58 -	.87	19.58 -	1.17	24.58 -	1.47
4.74 -	.28	9.74 -	.58	14.74 -	.88	19.74 -	1.18	24.74 -	1.48
4.91 -	.29	9.91 -	.59	14.91 -	.89	19.91 -	1.19	24.91 -	1.49

FEAST FOOD BAR
San Bruno, California

Guest Check   No.

Quantity		$	$
	Hamburger		
	Cheeseburger		
	Hot Dog		
	Small Pizza		
	Large Pizza		
	Sandwich		
	Submarine Sandwich		
	Soft Drink		
	Milk Shake		
	Coffee		
Guests	Attendant	Subtotal	
		Tax	
		Total	

FEAST FOOD BAR
San Bruno, California

Guest Check   No. 710   Date 4-15

Quantity		$	$
	Hamburger		
	Cheeseburger		
2	Hot Dog	.90	1.80
	Small Pizza		
	Large Pizza		
1	Sandwich *Cheese*		.85
	Submarine Sandwich		
2	Soft Drink	.35	.70
	Milk Shake		
1	Coffee		.40
Guests	Attendant	Subtotal	3.75
3	E. M.	Tax	.23
		Total	3.98

PROBLEMS

Here are some problems for practice in completing preprinted guest checks.

To do these problems, either (1) make copies of the unused preprinted guest check, or (2) use a blank piece of paper in place of a preprinted guest check form. Write down the check number and all the figures you would write on a guest check as shown in the illustration on page 170.

Complete guest checks for the following orders. Use your own initials on the check. Find the sales tax by using the tax table on page 170.*

1. Guest check No. 711, 2 people ordered:
    1 hamburger @ $1.25
    1 hot dog @ 90¢
    2 soft drinks @ 35¢ each

2. Guest check No. 712, 1 person ordered:
    1 cheeseburger @ $1.40
    1 soft drink @ 35¢

3. Guest check No. 713, 3 people ordered:
    2 hot dogs @ 90¢ each
    1 large pizza @ $2.25
    1 submarine sandwich @ $1.10
    2 soft drinks @ 35¢ each
    1 coffee @ 40¢

4. Guest check No. 714, 2 people ordered:
    1 small pizza @ $1.25
    1 sandwich, baloney @ 85¢
    2 coffees @ 40¢ each

5. Guest check No. 715, 4 people ordered:
    2 cheeseburgers @ $1.40
    1 hot dog @ 90¢
    1 sandwich, cheese @ 85¢
    3 coffees @ 40¢
    1 soft drink @ 35¢

6. Guest check No. 716, 2 people ordered:
    2 hamburgers @ $1.25
    2 milk shakes @ 75¢

7. Guest check No. 717, 1 person ordered:
    1 hot dog @ 90¢

8. Guest check No. 718, 3 people ordered:
    1 hamburger @ $1.25
    1 small pizza @ $1.25
    2 hot dogs @ 90¢
    3 soft drinks @ 35¢

9. Guest check No. 719, 3 people ordered:
    3 hamburgers @ $1.25
    2 milk shakes @ 75¢
    1 coffee @ 40¢

* Sales tax rates vary in different parts of the United States. The tax table used with these problems shows a tax rate of 6%.

10. Guest check No. 720, 4 people ordered:
    3 hot dogs @ 90¢
    1 cheeseburger @ $1.40
    1 sandwich, cheese @ 85¢
    4 soft drinks @ 35¢

11. Guest check No. 721, 2 people ordered:
    1 sandwich, pressed ham @ $1.00
    1 submarine sandwich @ $1.10
    2 soft drinks @ 35¢ each

12. Guest check No. 722, 1 person ordered:
    1 small pizza @ $1.25
    1 coffee @ 40¢

13. Guest check No. 723, 3 people ordered:
    3 cheeseburgers @ $1.40
    2 soft drinks @ 35¢
    1 coffee @ 40¢

14. Guest check No. 724, 4 people ordered:
    2 hamburgers $1.25
    2 hot dogs @ 90¢
    3 milk shakes @ 75¢
    1 coffee @ 40¢

15. Guest check No. 725, 3 people ordered:
    4 hot dogs @ 90¢
    2 soft drinks @ 35¢
    1 milk shake @ 75¢
    1 coffee @ 40¢

HANDWRITTEN

A great many restaurants do not use preprinted guest checks, especially those establishments that feature a genuine dining room atmosphere and personal service. Blank guest checks are handwritten in this type of restaurant.

Each restaurant has its own standard procedure for waiters to follow. In some places waiters write the guest check as they take the orders from the guests. In other places waiters first take the orders down quickly on a scratch pad as they are given, then later, when all ordering for that meal is completed, the waiter makes out a neat check to present to the guests. Sometimes there is an adding machine available to use in totalling the check. More often, however, the waiter/waitress must do his/her own arithmetic.

Some restaurants feature two kinds of menus, an a la carte menu and a table d'hote, or complete dinner, menu.

A la carte Menu. When ordering from the a la carte menu, the guest is charged for each item separately; for example $1.50 for a salad, $1.00 for soup, $7.00 for a steak and potato, $1.25 for a vegetable, $1.00 for pie, and 40¢ for coffee.

Table d'hote Menu. When ordering from the table d'hote (complete dinner) menu, the guest pays the amount shown opposite the main course, and the other courses are included free with this price. For example, if the price on the menu opposite steak is $8.25, this includes soup or salad (sometimes both), a vegetable, bread and butter, dessert, and a beverage, such as coffee or tea.

Cover Charge. Some restaurants add a charge, such as $2 or more, to each guest's check to help pay for live music and other entertainment provided. This is known as a cover charge.

Minimum Charge. Many restaurants show a minimum charge, such as $1 or $2, on their menus. This means that each guest will be charged at least that amount even if he or she orders something that costs less, such as a cup of coffee or a Coke only. The reason for this is that labor, rent, and other overhead expenses are so high, the restaurant cannot afford to serve a person food or beverage that sells for less than the minimum.

Writing Down the Order. As the waiter or waitress is taking the order, he or she must know whether the guest is ordering from the a la carte menu or the complete dinner menu. The waiter or waitress must write down the date, the number of people served, and his or her initials, as on the preprinted checks, plus the following information:

a. The entree requested.

b. How the meat is to be served—rare, medium, well done, etc.

c. Whether the guests want soup or salad, if there is a choice.

d. What kind of dressing they want on their salad.

e. What they want to drink.

f. Which dessert they want, if there is a choice.

Charging Correct Prices. Waiters get the prices for the guest checks from the menu. If they have been working there for quite a while, they may have most of the prices memorized. If a sales tax is involved, they will have a tax table to refer to.

Collecting a Tip. Usually, the tip for the waiter is not shown on the guest check. The guests decide how much tip to leave according to custom and how good the service was. However, in a few dining rooms, for example at certain clubs where members sign their meal checks and pay at the end of the month instead of paying cash, an amount may be added to the check for the tip based on a fixed rate, such as 10% or 15% of the total check.

On the following pages are some problems for practice in completing guest checks that are not preprinted. On these checks, the only thing that is preprinted is the check number; the waiter is required to write down the entire order and prices and total the check.

Pictured on page 174 is a completed guest check. The food and beverages listed on it can be found on the menu pictured on page 175.

According to this check, there were three people served by the waiter at one table. That is what is meant by "No. in Party."

Two people ordered from the complete dinner side of the menu. Their two orders of breaded veal cutlets cost $6.25 each for the complete dinner, $12.50 for the two.

No prices are shown opposite the salads, the two coffees, the vanilla ice cream, or the sherbet, as these are all included in the $12.50 dinner price.

```
                                        No. 800
          FEAST RESTAURANT
           San Bruno, California

  | No. in Party |   Waiter   |   Date   |
  |      3       |    E. M.   |   4-16   |
  |  2  | B. V. Cutlets - Din. 6.25 | 12.50 |
  |  2  | Salad T. I.              |       |
  |  2  | Cof.                     |       |
  |  1  | Van. I. C.               |       |
  |  1  | Sherb.                   |       |
  |     |                          |       |
  |  1  | N. Y. Steak M/R          |  7.00 |
  |  1  | Cof.                     |   .40 |
  |  1  | Pie - ap.                |  1.00 |
  |     |                          |       |
  |  1  | Burgundy Sm.             |  2.50 |
  |     |                          |       |
  |            Subtotal            | 23.40 |
  |              Tax               |  1.40 |
  |             Total              | 24.80 |

  Please pay waiter
```

A vegetable is also served on the plate with each main course ordered from the complete dinner menu.

The third person ordered from the a la carte section, so he was charged for each item separately.

In this restaurant, the waiters are permitted to abbreviate words when writing the check—as long as the order is clear. The following abbreviations were used:

B. V. for Breaded Veal
Din for Dinner
T. I. for Thousand Island dressing
Sherb for sherbet
Cof for coffee
N. Y. for New York
M/R for medium rare
Ap for apple
Sm for small
Van I. C. for vanilla ice cream

Feast Restaurant

* * * *

A La Carte Specials

Chicken Soup	$1.00
Vegetable Soup	1.00

- - -

Half Spring Chicken	4.25
New Orleans Prawns	4.75
Breaded Veal Cutlets	5.00
Ground Top Sirloin	4.75
New York Steak a la Minute	7.00
Calves Liver, Onions	5.50
Chicken Liver Saute	5.50
Roast Prime Ribs of Beef	7.75
Veal Scallopini	6.25

Served with Potato and Garnish

- - -

Spaghetti and Meat Balls	4.00
Spaghetti or Ravioli	3.25
Small Chef's Salad	1.50
Large Chef's Salad	2.50

- - -

Fresh Vegetables, Your Choice	1.25

- - -

Coffee, Tea, Milk	.40

- - -

Pie	1.00
Ice Cream	.90

Complete Dinners

Choice of Salad or Soup

Ground Top Sirloin	$6.00
Breaded Veal Cutlets	6.25
Half Spring Chicken	5.50
Broiled or Fried Fish—	
Salmon, Sole, Prawns	6.00
Abalone	8.50
Lobster Tails	9.25
Double Rib Lamb Chops	8.75
Calves Liver, Onions	6.75
Scallopini of Veal	7.50
New York Steak a la Minute	8.25
Roast Prime Ribs of Beef	9.00
Chicken Liver Saute	6.75
New York Cut Steak	9.75
Filet Mignon	10.00

Vegetable and potato
served with all entrees
Ice Cream or Sherbet
Coffee   Milk   Tea

Let us serve your next club dinner or luncheon. Ask our hostess for suggested menus.

WINE LIST

Red Wine	Small	Large	White Wine	Small	Large
Burgundy	*$2.50*	*$4.50*	*Sauterne*	*$2.50*	*$4.50*
Vin Rosé	*2.50*	*4.50*	*Chablis*	*2.50*	*4.50*

PROBLEMS

To write guest checks for each of the following orders, proceed as follows:

a. Read the order carefully.

b. Write on the check the check number, the number in the party, and your initials.

c. Notice whether the order is from the a la carte part of the menu or from the dinner section.

d. Write on the check each item ordered and how many.

e. Find the proper prices on the menu on page 175, and write them on the check. Be sure to double or triple the price if two or three portions of the item were ordered.
(Remember—there is no charge for soup or salad or dessert or the beverage with a complete dinner.)

f. Add the check to find the subtotal.

g. Find the proper tax on the "Sales Tax Collection Chart."

h. Add the tax to the subtotal and write in the total.

If you have no forms similar to the blank guest check shown here on which to do the problems, trace copies of this one or make copies on binder paper.

```
                              No. 801
        FEAST RESTAURANT
           San Bruno, California

   No. in Party    Waiter      Date




                          Subtotal
                          Tax

                          Total

   Please pay waiter
```

PROBLEMS

1. Guest check No. 801, 1 person ordered:
    Half spring chicken on the dinner menu
    Salad with French dressing
    Vanilla ice cream
    Coffee
    Small bottle of Chablis wine

2. Guest check No. 802, 1 person ordered:
    Ground top sirloin steak from the a la carte menu, cooked medium
    Large chef's salad with Roquefort dressing
    Apple pie
    Coffee

3. Guest check No. 803, 2 people ordered:
    2 New York cut steaks from the dinner menu
    1 soup
    1 salad with Thousand Island dressing
    1 vanilla ice cream
    1 sherbet
    2 coffees

4. Guest check No. 804, 2 people ordered:
    Double rib lamb chops from the dinner menu, cooked well
    Soup (on the dinner)
    Chocolate ice cream (on the dinner)
    Spaghetti and meat balls, a la carte
    Vanilla ice cream, a la carte
    2 coffees (1 is on the dinner)

5. Guest check No. 805, 3 people ordered:
    Veal scallopini from the dinner menu
    Liver and onions from the dinner menu
    1 soup (on the dinner)
    1 salad with French dressing (on the dinner)
    Prawns from the a la carte menu
    Large chef's salad with Roquefort dressing, a la carte
    2 vanilla ice creams (on the dinner)
    3 coffees (2 coffees are on the dinner)

6. Guest check No. 806, 2 people ordered:
    Roast prime rib of beef, rare, from the a la carte menu
    An order of fresh peas
    New York minute steak, medium rare, from the a la carte menu
    An order of fresh spinach
    Small bottle of burgundy wine
    2 coffees

7. Guest check No. 807, 3 people ordered:
    2 half spring chickens from the dinner menu
    1 filet mignon, medium rare, from the dinner menu
    1 soup
    2 salads, 1 with Thousand Island dressing, 1 with Roquefort
    2 vanilla ice creams
    1 sherbet
    2 coffees
    1 tea

8. Guest check No. 808, 3 people ordered:
   Abalone from the dinner menu
   Soup with the abalone
   Lobster tails from the dinner menu
   Salad, Roquefort dressing, with the lobster tails
   Breaded veal cutlets from the a la carte menu
   Chicken soup with the breaded veal cutlets (a la carte)
   2 coffees (on the dinner)
   1 milk, a la carte

9. Guest check No. 809, 4 people ordered:
   Ground top sirloin, cooked medium, from the dinner menu
   Prawns from the dinner menu
   Chicken liver saute from the dinner menu
   New Orleans prawns from the a la carte menu
   2 salads with French dressing, with the dinners
   1 soup, with the dinner
   1 order of fresh peas, a la carte
   3 vanilla ice creams, with the dinners
   3 coffees, with the dinners

10. Guest check No. 810, 2 people ordered:
    2 prime ribs of beef, 1 rare and 1 well done, from the a la carte menu
    1 chicken soup
    1 large chef's salad
    1 small bottle of vin rose wine
    2 vanilla ice creams
    2 coffees

11. Guest check No. 811, 3 people ordered:
    2 New York minute steaks, 1 medium rare and 1 rare, from the dinner menu
    2 salads, 1 with Roquefort dressing and 1 with French dressing, with the steaks
    1 order of ravioli, a la carte
    A large bottle of burgundy wine
    2 vanilla ice creams with the steak dinners
    3 coffees (2 are with the dinners)

12. Guest check No. 812, 2 people ordered:
    2 New York cut steaks, 1 rare and 1 medium rare, from the dinner menu
    2 soups
    2 sherbets
    2 coffees

QUESTIONS

1. What type of restaurants use preprinted guest checks?
2. Why are guest checks for higher-priced restaurants not preprinted?
3. List at least three important points to remember when filling in guest checks.
4. What is meant by ordering "a la carte?"
5. What is meant by "table d'hote" menu?

CASH RECEIPTS REPORT

Cashiers in restaurants, hotels, or any other business usually fill out a cash receipts report when they complete their work shift and hand the money in to the auditor or manager.

The report will be similar to the one shown on this page, and it makes the counting of the cash very easy.

Notice that when businessmen use the word "cash" they mean checks and paper money as well as coin. Paper money is called "currency" on the Cash Receipts Report.

The main reason for filling out the report is to find out if the money the cashier has in his/her cash drawer agrees with the total amount of cash sales made during his/her shift. This is known as "proving cash" or "balancing cash" at the end of the shift.

To fill out the Cash Receipts Report:

a. Under "Number" you enter the number of pennies, nickels, dimes, $1 bills, $5 bills, etc., that are in the cash drawer.

b. Multiply the various numbers by their denominations to get the Amount of each kind.

c. Add the Total Coin, Total Currency, and Total Checks to find the Total Cash in Drawer.

d. Continue filling in the remainder of the report as it indicates.

Cash Paid Out is found by adding the receipts in the drawer for small amounts that were paid out of the cash register during the shift.

The cashier starts his/her shift with a certain amount of change, for example, $50. This is entered on the report where it says "Less Change."

The Cash Sales total is found on the cash register tape. This tape is also called a "detailed audit strip." Each sale that is rung up on the cash register is automatically printed and totalled on the tape.

Total cash sales in a restaurant can also be found by adding up all the waiters' guest checks during that shift.

QUESTIONS

1. What does a businessman mean by the word "cash?"
2. What is "currency?"
3. What is meant by "proving cash?"
4. How do you know how much cash has been paid out during the shift?
5. Where do you find the total cash sales for the shift?

CASH RECEIPTS REPORT

Date: 6/14/— Cashier: B. C. Jones

Number	Denomination	Amount
	Coin:	
37 @	1¢	0.37
63 @	5¢	3.15
48 @	10¢	4.80
26 @	25¢	6.50
7 @	50¢	3.50
2 @	$1.00	2.00
	Total Coin:	20.32
	Currency:	
38 @	$1	38.00
24 @	$5	120.00
18 @	$10	180.00
15 @	$20	300.00
	Other	

Total Currency 638.00
Total Amount of Checks 15.00
Total Cash in Drawer 673.32
Add Cash Paid Out 3.00
Total Cash 676.32
Less Change 50.00
Cash Received, Actual 626.32
Cash Sales 626.35
Amount Short or Over −.03

PROBLEMS

Make up five forms for Cash Receipts Reports like the one shown here.

Fill out the reports from the information given below. Use the date shown. Sign your name on the form where it says "Cashier."

June 15, 19___

52 pennies   38 $1 bills
29 nickels   16 $5 bills
35 dimes   13 $10 bills
17 quarters   12 $20 bills
 9 half dollars   1 check for $10
Cash paid out, $5. Starting change, $40.
Cash sales rung up according to cash register tape, $477.22.

June 16, 19___

27 pennies   41 $1 bills
38 nickels   27 $5 bills
43 dimes   15 $10 bills
19 quarters   14 $20 bills
11 half dollars   2 checks: $10 & $25
 1 silver dollar
Cash paid out, $7.50. Starting change, $50.
Cash sales rung up according to cash register tape, $616.27.

June 17, 19___

19 pennies   47 $1 bills
65 nickels   32 $5 bills
39 dimes   19 $10 bills
28 quarters   17 $20 bills
 7 half dollars   2 checks: $10 & $15
Cash paid out, $6.50. Starting change, $60.
Cash sales according to cash register tape, $726.44.

June 18, 19___

48 pennies   28 $1 bills
43 nickels   24 $5 bills
29 dimes   23 $10 bills
22 quarters   19 $20 bills
 8 half dollars   3 checks @ $10
No cash paid out. Starting change, $50.
Cash sales according to cash register tape, $753.03.

June 19, 19___

33 pennies   53 $1 bills
58 nickels   26 $5 bills
16 dimes   18 $10 bills
29 quarters   21 $20 bills
 5 half dollars   1 check @ $5.50
Cash paid out, $3. Starting change, $60.
Cash sales according to cash register tape, $846.33.

CASH RECEIPTS REPORT

Date _____ Cashier _____

Number	Denomination	Amount
	Coin:	
@	1¢	
@	5¢	
@	10¢	
@	25¢	
@	50¢	
@	$1.00	

Total Coin: _____

Currency:
@	$1	
@	$5	
@	$10	
@	$20	
	Other	

Total Currency: _____
Total Amount of Checks _____
Total Cash in Drawer _____
Add Cash Paid Out _____
Total Cash _____
Less Change _____
Cash Received, Actual _____
Cash Sales _____
Amount Short or Over _____

Permission is granted to teachers to reproduce this form.

BUSINESS FORMS

Tons of paper are necessary to keep track of the business that goes on every day in the United States. In fact, that is the reason business organizations are using data processing and computers more and more; they cannot keep up with the paper work, and they need electronic data processing methods to help speed up the flow of information.

Most of the paper work is handled by office workers, who are also called "white collar workers." In addition to the people keeping track of business in offices, however, truck drivers, storeroom clerks, cooks, waiters, hotel desk clerks, room maids, and countless other people who are not strictly "office workers" must use business forms in connection with their work. There are very few jobs that do not require the employee to keep some simple record, or count something, or sign a receipt, or write something down.

REQUISITIONS, PURCHASE ORDERS, AND INVOICES

In addition to checks, which are used for money, three of the most common business forms, or pieces of paper, that are used are requisitions, purchase orders, and invoices. There are many other business forms, such as sales slips, receipts, payroll records, inventory records, but requisitions, purchase orders, and invoices are among the most basic, and everyone working in the hotel and restaurant business, in particular, should be familiar with them.

Requisitions, purchase orders, and invoices are used in any business that buys and sells. The flow chart on the following page shows the place of these three forms in the movement of food and beverage supplies from the wholesale food supply company to the restaurant storeroom and from the storeroom to the kitchen.

Requisitions

When a cook needs some cans of tuna, or coffee, or ten heads of lettuce, or some mustard, or anything else to prepare his meals, he fills out a requisition, such as the one pictured on page 183 and takes it to the storeroom, where a supply of various foods and beverages is kept.

The requisition lists the amount and description of each item taken from the storeroom. It may show the price of each individual item and the extensions and total value of the items being taken out; or it may not show any prices, depending on the system used in that storeroom.

The requisition is signed by the person who is going to use the items. It may also require the signature of someone else, such as the chef, or the manager, as a safeguard against theft or waste.

The requisition is actually a receipt. It is given to the storeroom keeper in exchange for the items taken out. The reasons for using requisitions are:

FLOW CHART SHOWING USE OF REQUISITIONS, PURCHASE ORDERS, AND INVOICES

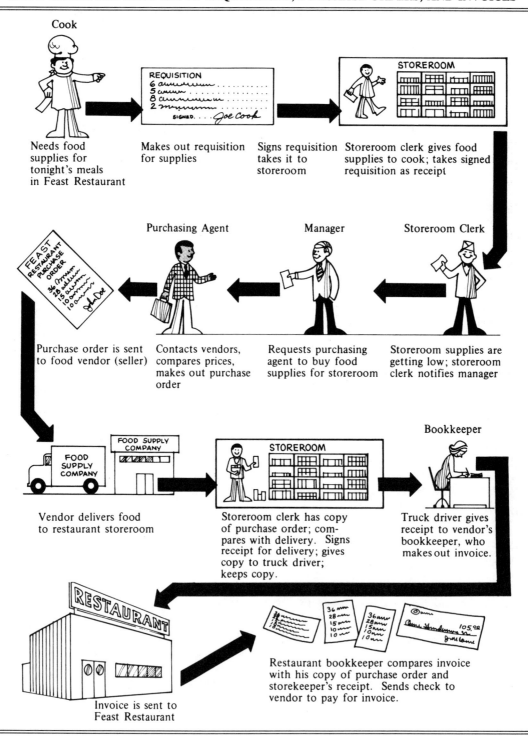

1. To account for the food and beverage issued from the storeroom each day.
2. To insure that the food and beverage items are issued only to the proper persons.

Thus, at the end of the day, the storeroom clerk must have requisitions for everything that has been taken from the storeroom.

STOREROOM REQUISITION

FEAST RESTAURANTS
San Francisco, California

Storeroom Requisition

Charge To __Kitchen__   Date __6-15__

Amount	Unit	Item	Unit Price	Extension
3	cn.	Tuna, chunks, 66 oz.	$4.46	$13.38
2	cn.	Coffee, regular, 2 lb.	4.18	8.36
1	gal.	Mustard, prepared	1.46	1.46

Total __$23.20__

Signed __D. J. Golladay__
Approved __K. Grimes__

Purchase Orders

A purchase order is a business form which asks a vendor (seller) to deliver a certain number of items at a certain price to a certain place. A purchase order is signed by someone such as a purchasing agent, or a chef or a manager; and it means that if the proper merchandise is delivered, as listed on the purchase order, the buyer will pay for it.

The person doing the buying compares prices and quality among a number of vendors and then gives the purchase order to the one offering the best deal.

The original copy of the purchase order is sent to the vendor as authority to deliver the merchandise, and carbon copies are kept by the restaurant or hotel buying the merchandise so that the storeroom clerk and the person who pays the bills will know what was ordered.

Purchase orders eliminate arguments about how many items were ordered, or what was ordered, or who ordered the merchandise. Without purchase orders, someone might order some steaks or wine, for example, and charge it to a restaurant whose manager knew nothing about it.

PURCHASE ORDER

FEAST RESTAURANTS 127 Army Street San Francisco, California			PURCHASE ORDER No. 1293 Date: July 1, 19	
To:	Millbrae Food Supply Co. 1295 Broadway Millbrae, California	Ship to:	Feast Restaurants Storeroom 127 Army Street San Francisco, California	
		Delivery Date:	No later than 7-10	

Please deliver the following goods, which are purchased in accordance with the prices and descriptions given below. Show our purchase order number on all deliveries and invoices.

Description	Unit	Quantity	Unit Price	Amount
Tuna, light meat, chunks, 6/66's	cs.	10	$24.60	$246.00
Coffee, regular grind, Acme, 12/2 lb.	cs.	15	44.00	660.00
Powdered sugar, 25 lb.	pkg.	4	7.40	29.60
Total				$935.60

M. J. Kinnealy
Purchasing Agent

Purchase Specifications

The vendor should get a copy of the buyer's purchase specifications with the purchase order. Sometimes the specifications are written on the purchase order itself.

Purchase specifications are detailed descriptions of the food or beverage being bought. If grapefruit is being bought, for example, the specifications might describe the size and weight of each grapefruit, the color, the softness, and perhaps the brand, as well as the number requested.

A copy of the purchase specifications is given to the storeroom clerk so that when the goods are received, he or she can check to see that they are exactly what was ordered.

Invoices

An invoice is simply a request for payment for goods sent by a vendor to a purchaser. Each company has its own design for its invoices. Some are very plain and some are fancy. Some are blank forms that are filled in with names and prices of the items sold. Other invoices have printed on them all the items that company usually sells, and the number of each item sold and the prices are filled in for this particular purchase.

Most invoices, however, show the following things:

1. The name and address of the vendor (seller).

2. The name and address of the purchaser.

3. An invoice number.

STANDARD PURCHASE SPECIFICATIONS FOR THE EXAMPLE RESTAURANT*

SIRLOIN STRIP STEAKS—14 ounces (bone-in) Top. Choice.

1. Sirloin strips from which steaks are cut shall be:
   1.1 Bone-in with medium size eye
   1.2 Have chine bone trimmed close to the meat
   1.3 Aged 2½ to 4 weeks
2. Individual steak shall:
   2.1 Weigh 14 ounces (bone-in) with maximum allowance of 1/2 ounce plus or minus
   2.2 Have no more than 1/2" tail or flank measured from the main muscle (eye)
   2.3 Have no more than 1/4" fat covering over the main muscle (eye)
   2.4 Include no nerve and steaks from the hip end
   2.5 Be properly packaged for refrigeration storage

BEEF RIBS—Oven Prepared; Top Choice

1. Ribs from which an oven-prepared rib is fabricated shall:
   1.1 Be cut from a 7 bone rib, to measure on the flank no more than 1½" from the meat of the main muscle (eye) on the loin end and no more than 4½" from the inside of the chine bone on the chuck end.
      1.1.1 Must be cut straight between these two points. Care should be taken to insure this straightness.
   1.2 Show white buttons between ribs, porous bones and clear, white fat.
   1.3 Chine bone remove squarely to where the meat splits.
   1.4 Ribs should not show heavy fat layers on chuck end or along flank, nor should the back have heavy fat covering.
   1.5 Ribs should be aged not less than 10 days nor more than 21 days from date of slaughter.
2. An oven-prepared rib shall:
   2.1 Have the cap removed the entire length of the rib, and the lean meat trimmed from the cap.
   2.2 Have the back strap, blade bone, and blade bone cartilage removed, and the back securely tied to its natural position.
   2.3 The fabricated rib should weigh not less than 19 lbs., nor more than 23 lbs.

LAMB FORE (CHUCK)

1. Double, including neck, shanks and briskets, and 1st through 4th ribs. (All in one piece.)
2. To weigh 16 to 18 lbs.

BROILERS. U.S. GOVT. GRADE A

1. Eviscerated to weigh 2 lbs., careful attention to be given to sizing within plus or minus 1/8 lb. weight tolerance. Total weight of a number of broilers must be within plus or minus 5%.

LOBSTERS

1. Live, to weigh 1½ lbs. careful attention to be given to sizing within plus or minus 1/8 lb. weight tolerance. Total weight of a number of lobsters must be within plus or minus 5%.

Source: Hotel Corporation of America. Taken or adapted from their set of Standard Purchase Specifications.

INVOICE

MILLBRAE FOOD SUPPLY CO.
1295 Broadway
Millbrae, California

INVOICE NO: 70-105

Date: 7-15

Sold to: Feast Restaurants
127 Army Street
San Francisco, California

Delivery Date	How Shipped	Your Purchase Order No.	Terms
7-10	Truck	1293	2/10 n/30

Quantity	Unit	Description	Unit Price	Total
10	cs.	Tuna, light meat, chunks, 6/66's	$24.60	$246.00
15	cs.	Coffee, regular grind, Acme, 12/2 lb.	44.00	660.00
4	pkg.	Powdered sugar, 24 lb.	7.40	29.60
		Total		$935.60

4. The date the invoice was sent.
5. The buyer's purchase order number.
6. How the goods were shipped (by truck, etc.).
7. When the goods were shipped.
8. When payment of the invoice is due.
9. Discount terms, if there are any.
10. What was shipped to the buyer.
11. How many of each item were shipped.
12. The price of each item and the total price.
13. Taxes and freight charges, if there are any.

Discount Terms

Two types of discounts that may be shown on invoices are quantity discounts and cash discounts.

A quantity discount means that the buyer is given a lower price because he is buying a large quantity. The quantity discount may be a deduction of anywhere from 10% up to 30% from the regular price.

A cash discount is sometimes given to the buyer for paying his bill quickly. For example, if the invoice shows terms of 2/10, n/30, that means 2% discount in 10 days, net 30 days. In other words, the buyer may either pay the invoice with 2% deducted from the total within 10 days of the date of the invoice or pay the full amount of the invoice within 30 days.

QUESTIONS

1. What is a requisition?
2. What is a purchase order?
3. What are purchase specifications?
4. What is an invoice?
5. What is a cash discount?
6. On an invoice what does the notation 2/10 n/30 mean?

PROBLEMS

Make up some invoice forms similar to the blank form shown on page 187 on which to do the following problems.

You work for the Millbrae Food Supply Co. Make out invoices for each of the following sales. Show the items and the prices and totals as they are shown in the illustration on page 185. Use today's date on each invoice. Show the date of one week ago as the delivery date.

MILLBRAE FOOD SUPPLY CO.				INVOICE NO:_____	
1295 Broadway					
Millbrae, California				Date:_____	
Sold to: _____					
_____					
_____					
Delivery Date	How Shipped		Your Purchase Order No.		Terms
Quantity	Unit	Description		Unit Price	Total

Permission is hereby granted to the teacher using this book to reproduce this page as needed.

Invoice No.

79–1 Sold to San Bruno Restaurants, 327 First Avenue, San Bruno, California, on terms of 2% discount if paid within 10 days of the date of the invoice, net 30 days, on their purchase order No. 1103:
  2 pkg. Brown sugar, 100 lb. @ $23.70 per pkg.
  2 pkg. Powdered sugar, 25 lb. @ $7.40 per pkg.
  3 cases Prepared mustard, 4/1 gal. @ $5.96 per case
  2 pkg. Lasagne, 10 lb. @ $3.92 per pkg.

79–2 Sold to San Francisco Restaurant, 805 Geary Street, San Francisco, California, on terms of net 30 days, on their purchase order No. 715:
  1 pkg. Onion powder, 1 lb. @ $1.96 per pkg.
  3 cases Beans, French cut, 24/303 @ $7.96 per case
  2 cases Catsup, medium weight, 6/10 @ $11.96 per case
  2 cases Eggs, dry whole, 6/10 @ $50.40 per case

Invoice No.

79–3   Sold to Joe's Restaurant, 212 Front Street, Sacramento, California, on terms of 2/10, net 30 days, on their purchase order No. 3–102:
   1 pkg. Noodles, lasagne, 10 lb. @ $3.90 per pkg.
   4 cases Peaches, cling, sliced, 6/10 @ 11.80 per case
   2 cases Relish, sweet, 4/1 gal. @ $11.90 per case
   6 cases Soup, tomato, 12/3 @ $13.70 per case
   2 pkg. Brown sugar, 100# @ $23.44 per pkg.

79–4   Sold to Far West Hotels, 721 Geary Street, San Francisco, California, on terms of 2/10, net 30, on their purchase order No. 6534:
   1 qt. Food coloring, red @ $2.20 per qt.
   6 pkg. Spaghetti, 20 lb. @ $5.10 per pkg.
   2 pkg. Garlic powder, T-180, 11 oz. @ $2.58 per pkg.

79–5   Sold to Sea King Restaurants, 302 Broadway, Millbrae, California, on terms of 1/10, net 30, on their purchase order No. 89–2:
   6 cases Tuna, solid, light meat, 6/66's @ $26.90 per case
   2 pkg. Salt, table, 25 lb. @ $1.90 per pkg.
   4 pkg. Italian seasoning, 6 oz. @ $2.10 per pkg.
   5 cases Coffee, reg., 12/2 lb @ $36.98 per case

79–6   Sold to Pacific Hotel, 605 Ocean Avenue, Pacifica, California, on terms of 2/10, net 30, on their purchase order No. 3225:
   12 gal. Vanilla, imi., ST-180 @ $6.20 per gal.
   3 cases Apple slices, water pack, 6/10 @ $12.90 per case
   5 cases Roast Beef, 6/6's @ $49.90 per case
   2 cases Bacon bits, 6/10 @ $19.30 per case

79–7   Sold to Seaview Restaurant, 305 Jones Street, Oakland, California, terms of net 30 days, their purchase order No. 407:
   3 pkg. Cinnamon, 6 lb @ $36.50 per pkg.
   2 cases Worcestershire sauce, 4/1 gal. @ $3.10
   5 cases Soup, vegetable, 12/3 @ $13.80 per case
   4 cases Tuna, light meat, chunks, 6/66's @ $24.60 per case

79–8   Sold to the Parkview Hotel, 702 Laurel Street, San Mateo, California, terms of 2/20, n/30, on their purchase order No. 1237:
   5 cases Beans, French cut, 24/303 @ $7.96 per case
   3 cases Eggs, dry whole, 6/10 @ $50.40 per case
   2 pkg. Sugar, granulated, 100 lb. @ $22.86 per pkg.

79–9   Sold to Bayview Retirement Hotel, 105 Scott Street, San Francisco, California, terms of net 30 days, their purchase order No. 726:
   4 cases Coffee, drip grind, 12/2 lb. @ $36.98 per case
   3 cases Bacon bits, 6/1 lb. @ $20.80 per case
   3 cases Apple slices, water packed, 6/10 @ $12.90 per case
   6 cases Peaches, cling, sliced, 6/10 @ $11.80 per case
   6 cases Soup, tomato, 12/3 @ $13.70 per case

Invoice No.

79–10 Sold to Uncle Pete's Restaurants, 201 Broadway, San Francisco, California, terms of 2/10, n/30, their purchase order No. 3–215:
    4 cases Lasagne, 10 lb. @ $3.92
    2 cases Salad vinegar, 4/1 gal. @ $5.70 per case
    2 pkg. Salt, table, 25 lb. @ $1.90 per pkg.
    3 jars Cherries, maraschino, 1/2's @ $5.98 per jar
    4 pkg. Italian seasoning, 6 oz. @ $2.10 per pkg.
    2 pkg. Allspice, ground S-3, 1 lb. @ $3.50 per pkg.

79–11 Sold to Cookie's Restaurant, 605 42nd Street, San Francisco, California, terms of net 30 days, their purchase order No. 2113:
    8 pkg. Spaghetti, broken, no stick, 20 lb. @ $5.10 per pkg.
    5 pkg. Italian seasoning, 6 oz. @ $2.10 per pkg.
    4 cases Roast beef, 6/6 lb. @ $49.90 per case
    2 cases Coffee, reg. grind, 12/2 lb. @ $36.98 per case
    3 cases Relish, hot dog, 4/1 gal. @ $13.50 per case

79–12 Sold to Bob's Place, 711 Marina Place, San Francisco, California, terms of 2/10, net 30 days, their purchase order No. 881:
    5 cases Tuna, light meat, solid, 6/66's @ $26.90 per case
    3 jars Cherries, maraschino, 1/2's @ $5.98 per jar
    3 cases Catsup, medium wt., 6/10 @ $11.96
    1 pkg. Pepper, black, S-192, 1 lb. @ $1.78 per pkg.

STOREROOM FUNDAMENTALS

The job of the storeroom clerk, sometimes called storekeeper, is extremely important. He/she is responsible for receiving and issuing foods and beverages usually worth thousands of dollars. Mistakes or thefts in this department can turn the organization's profit into a loss.

Storekeeper's Duties

In receiving the merchandise, usually from trucks, he or she must check such things as:

1. The quantity

2. The description (Is it exactly what was ordered?)

3. The quality (Is it in good condition?)

Usually the storekeeper will have some record, such as a copy of the purchase order and specifications, with which to compare the items received.

Once the merchandise is accepted by the storekeeper and the truck driver has been given a receipt, it becomes the property of the storekeeper's company and the company must pay for it. Therefore, if the merchandise is not in good condition, or the amount is not correct, the storekeeper must not accept it.

Physical Inventory

A physical inventory of storeroom items is taken once a month or more often. A physical inventory means actually counting everything in the storeroom.

Perpetual Inventory

A perpetual inventory is also kept daily by the storeroom clerk. A perpetual inventory is a record in a book or on cards which shows the balance on hand at any moment for each item in the storeroom. As an item is issued on a requisition, the amount is subtracted from the book inventory balance. As new items are received, the amounts on the receipts are added to the book inventory balance.

At the end of the month, the physical inventory should be the same as the book inventory. If they do not agree, the problem may be due to theft or some other problem which needs attention.

Par Stock

How much is kept on hand in storerooms? Persons involved with the use of the food and beverage supplies, such as the chef, the food and beverage controller, and perhaps the storekeeper decide what is the least amount of each item it is safe to have on hand to operate and the most that should ever be on hand in the storeroom. These minimums and maximums are known as par stock. Whenever an item gets down to its minimum, an order is put in to buy enough of the item to bring it back up to its maximum.

First In, First Out

Another principle of good storeroom procedure is to follow the "first in, first out" routine; that is, new items received are put on the back of the shelves, and older items are moved toward the front and issued first. In this way, nothing should get too old and spoil.

QUESTIONS

1. What is a physical inventory?
2. What is a perpetual inventory?
3. What is par stock?
4. What is meant by "first in, first out" in storeroom operation?

FINANCIAL STATEMENTS

All business organizations use two basic fundamental statements:

1. A balance sheet.
2. An income statement (sometimes called a profit and loss statement).

These statements are used by individual persons, as well as business organizations, to report their financial affairs to the government for income taxes, to get loans from banks and other places, to report on the progress of the business for partners or stockholders, and other activities.

BALANCE SHEET

A balance sheet shows how much a person or a company is worth on a particular date. It is simply a list written or typed on a piece of paper showing: (1) what a person or business owns—called assets; (2) what a person or business owes—called liabilities; and (3) how much a person or business is worth—called net worth, or proprietorship.

We figure net worth by subtracting the total amount owed from the total value of all the things owned. In other words:

assets − liabilities = net worth

A balance sheet for a person requesting a loan at a bank would look like this:

```
                        Carl A. Johnson
                         Balance Sheet
                        As of July 15, 19__
```

Assets		Liabilities	
Cash in Checking Account	$1,312.50	Mortgage on Home	$78,358.75
Cash in Savings Account	2,400.00	Auto Loan - Bank	4,150.58
U.S. Savings Bonds	375.00	Charge Accounts - Dept. Stores	327.72
Automobile	6,500.00	Total Liabilities	82,837.05
Home	98,000.00		
Furniture and Appliances	4,200.00	Net Worth	29,950.45
Total Assets	$112,787.50	Total Liabilities and Net Worth	$112,787.50

Finding Net Worth

When Carl A. Johnson made out his balance sheet, he found that the total value of all his assets was $112,787.50. His liabilities totalled $82,837.05. He subtracted the liabilities from the assets and found that his net worth was $29,950.45.

In writing down his balance sheet items, Carl A. Johnson showed his assets on the left side of the sheet. On the right side he listed his liabilities. When he found his net worth, he wrote that figure below his liabilities.

To prove that he had figured his net worth correctly, he then added the total liabilities and net worth. This answer, $112,787.50, should be the same as the total assets.

Assets Equal Liabilities Plus Net Worth

The total assets always equal, or balance, the total of the liabilities and the net worth. That is why this financial statement is called a "balance sheet."

A balance sheet for the City Restaurant, which is owned by John Smith, would look like this:

City Restaurant
Balance Sheet
As of December 31, 19—

Assets		Liabilities	
Cash	$ 7,100.00	Notes Payable – Bank	$ 5,200.00
Accounts Receivable	1,722.50	Accounts Payable	1,472.25
Food & Beverage Inventory	3,228.00	Payroll Taxes Payable	432.50
Supplies	781.25	Total Liabilities	7,104.75
Equipment	4,678.00		
Furniture & Fixtures	2,171.00	Net Worth	
		John Smith, Capital	12,576.00
Total Assets	$19,680.75	Total Liabilities and Net Worth	$19,680.75

Business Vocabulary

In the balance sheet shown above, "Cash" means all the cash the City Restaurant has in the bank and on hand in the cash register on December 31. "Accounts Receivable" means money that is owed to the restaurant by customers. "Accounts Payable" means money that the restaurant owes to people it buys food and beverages and supplies from.

The word "Capital" after John Smith's name shows the value of the owner's investment in the business, another word for his "net worth" in the restaurant.

PROBLEMS

Prepare a balance sheet for each of the following sets of facts. In each case, you will find the net worth by subtracting the total liabilities from the total assets. When you finish the balance sheet, it should look like the ones shown on the previous page.

1. Mr. Ken LaCrosse, on June 15 of the current year, had the following assets: Cash, $2,265.50; U. S. Savings Bonds, $750.00; Automobile, $1,235.00; Furniture and Appliances, $900.00. His liabilities were: Note Payable, Bank, $825.00; Accounts Payable, Stores, $123.50.

2. On September 30 of the current year the Peninsula Restaurant had the following assets: Cash $3,278.90; Accounts Receivable, $732.50; Food and Beverage Inventory, $2,156.25; Supplies, $381.25; Equipment, $3,980.00; Furniture and Fixtures, $6,229.85. The liabilities of the restaurant were: Note Payable, Bank, $4,150.00; Accounts Payable, $690.00; Payroll Taxes Payable, $342.80. The restaurant is owned by Myrna Dake.

3. On May 30 of the current year Joe's Hamburger Stand had the following assets: Cash $876.35; Food and Beverage Inventory, $926.50; Supplies, $78.50; Equipment, $237.90; Furniture and Fixtures, $290.00; Rent Deposit, $100.00. Liabilities for the hamburger stand were: Note Payable to Arnold Jones, $825.00; Accounts Payable, $76.40; Sales Tax Payable, $58.22. The hamburger stand is owned by Joe Johnson.

4. James and Jennifer Jermyn had the following assets on January 15 of this year: Cash, $3,225.00; Stock, General Motors, $550.00; Automobile, $8,900.00; Home, $132,500.00; Furniture and Appliances, $4,850.00. Their liabilities were: Note Payable, Credit Union, $1,468.00; Mortgage Payable on Home, $86,742.40; Charge Accounts, Department Stores, $376.50.

5. John's Drive-Inn had the following assets on June 30 of this year: Cash, $2,168.90; Food and Beverage Inventory, $2,152.35; Supplies, $427.85; Equipment, $6,380.00; Furniture and Fixtures, $3,895.00; Rent Deposit, $175.00; Utilities Deposits, $45.00. Liabilities of the drive-inn were: Note Payable, Equipment Co., $3,480.00; Accounts Payable, $387.25; Payroll Taxes Payable, $128.20; Sales Tax Payable, $386.78. John's Drive-Inn is owned by James and Maureen Smith.

6. The Ocean View Restaurant had the following assets on December 31 of last year: Cash, $3,288.50; Accounts Receivable, $1,428.50; Food & Beverage Inventory, $4,621.25; Supplies, $689.20; Equipment, $6,214.75; Furniture and Equipment, $10,820.00; Building $42,500.00; Land, $20,000.00. The restaurant's liabilities were: Mortgage Payable, Bank, $38,450.00; Accounts Payable, $866.42; Payroll Taxes Payable, $376.90; Sales Taxes Payable, $628.30. The Ocean View Restaurant is owned by Melville Cook.

INCOME STATEMENT

An income statement is a report which shows the results of the operation of a business for a certain length of time; for example, for one month, for three months, or for a year. The owner of a business can make up an income statement whenever he or she wants to know how much money the business is making or how much it might be losing.

Information from the Set of Books

The owner gets the figures to make the statement from the daily records kept by the bookkeeper. This daily record of the business is called the "set of books." Every business must have a set of books for income tax purposes.

The statements of income for all businesses that sell goods of any kind, including food and beverages, are very similar. They show the three things the owners are most interested in:

1. The total income—all the money taken in.

2. The gross profit—the total income less the cost of the goods that were sold.

3. The net profit, or net income—the profit that remains after all expenses have been paid.*

On the next page is an illustration of an income statement for a restaurant. It shows the results of operating for a period of one year.

Notice that there are two columns for money figures. The first column on the left shows the details, and the second column is used only for totals. (There is also a narrow, unused column remaining on the right edge of the statement. This space will be explained and used in the next lesson.)

PROBLEMS

Prepare a statement of income for each of the following sets of facts.†
Use the illustration on the following page for a model.

1. A restaurant named the San Mateo Buffet had sales for the month of October of $40,000. On October 1, its inventory of food and beverages was $6,250. During the month it purchased $18,000 worth of food and beverages. At the end of October its inventory was $8,250.

    Expenses of the restaurant for the month were: salaries, $10,000; payroll taxes, $1,000; employees' welfare and pensions, $1,600; taxes and licenses, $120; utilities, $240; rent, $1,200; insurance, $80; laundry, $320; depreciation, $200; supplies, $600; telephone, $160; accounting, $80; miscellaneous, $40.

2. Carl's Drive-In restaurant had sales for the month of January of $50,000. On January 1, its inventory of food and beverages was $6,790. During the month it had purchases of food and beverages amounting to $21,200. At the end of January its food and beverage inventory was $5,490.

    Expenses for the month were: salaries, $13,250; payroll taxes, $1,350; employees' welfare and pensions, $2,000; taxes and licenses, $150; utilities, $250; rent, $2,050; insurance, $200; laundry, $400; repairs and maintenance, $165; depreciation, $285;

---

* The word *gross* means total, before deductions; hence, gross profit means profit on selling before deducting expenses. The word *net* means after all deductions; hence, net income, or net profit, means income after all expenses have been deducted.

† Save the completed income statements made for the problems above; they will be used again in the next lesson.

supplies, $350; telephone, $152; accounting, $100; advertising, $830; and miscellaneous, $170.

3. The Oceanside Restaurant had sales for the year ending December 31 of $360,000. The beginning inventory on January 1 was $15,736. During the year it purchased $171,409 worth of food and beverages. At the end of the year its inventory was $17,225.

   The restaurant had expenses during the year as follows: salaries, $100,008; payroll taxes, $10,800; employees' welfare and pensions, $14,400; taxes and licenses, $1,080; utilities, $1,885; rent, $18,000; insurance, $1,764; laundry, $2,520; repairs and maintenance, $720; depreciation, $3,600; supplies, $1,445; telephone, $1,490; loan interest, $30,000; legal and accounting, $1,200; advertising, $2,400; miscellaneous, $329.

4. Joe's Lounge had sales for the month of March of $31,000. On March 1 its inventory of food and beverages was $2,758. On

Millbrae House Restaurant
Income Statement
For Year Ending December 31, 19—

Income			
Sales			$332,450
Cost of Food & Beverages Sold:			
Food & Bev. Inventory, Jan. 1		$17,692	
Purchases		145,898	
Total Food & Bev. Available for Sale		163,590	
Less Food & Bev. Inventory, Dec. 31		−18,975	
Cost of Food & Bev. Sold		144,615	−144,615
Gross Profit			187,835
Expenses:			
Salaries		86,440	
Payroll Taxes		8,652	
Employee Welfare & Pensions		13,830	
Taxes & Licenses		665	
Utilities		1,427	
Rent		12,000	
Insurance		1,329	
Laundry		2,327	
Repairs & Maintenance		1,662	
Depreciation		3,325	
Supplies		665	
Telephone		997	
Loan Interest		1,330	
Legal & Accounting		1,325	
Advertising		2,000	
Miscellaneous		1,650	
Total Expenses		139,624	−139,624
Net Income or <Loss>			$48,211

Note: In the illustration above and in the problems in this section, no cents amounts are used. All figures have been rounded off to even dollars. Businessmen frequently do this in making up financial statements to make them easier to read.

March 31 its inventory was $2,728. Its purchases for the month were $13,300.

Expenses for the month were: salaries, $8,224; payroll taxes, $832; employees' welfare and pensions, $986; taxes and licenses, $74; utilities, $298; rent, $1,500; insurance, $125; laundry, $186; repairs and maintenance, $65; depreciation, $136; supplies, $110; telephone, $279; accounting, $80; miscellaneous, $63.

5. Mom and Dad's Diner had sales for the year of $285,000. Its food and beverage inventory on January 1 was $7,348. Its purchases for the year amounted to $135,533. The inventory on December 31 was $7,221.

Expenses for the year were: salaries, $62,700; payroll taxes, $6,350; employees' welfare and pensions, $9,725; taxes and licenses, $730; utilities, $1,732; rent, $11,525; insurance, $1,440; laundry, $2,520; repairs and maintenance, $285; depreciation, $3,000; supplies, $855; telephone, $1,425; accounting, $1,200; miscellaneous, $286.

6. The White Cow Creamery had sales for the month of July of $28,400. The creamery's inventory of food and beverages on July 1 was $1,520. On July 31 it was $1,968. Purchases for the month were $14,080.

Expenses were: salaries, $5,822; payroll taxes, $640; employees' welfare and pensions, $960; taxes and licenses, $115; utilities, $220; rent, $1,200; insurance, $142; laundry, $57; depreciation, $285; supplies, $171; telephone, $59; accounting, $57; advertising, $120; miscellaneous, $29.

Figuring Percentages of Sales

Figuring percent of sales is very easy. We simply divide the sales figure into the other number. In the income statement for the Millbrae House Restaurant, shown in the illustration on page 198, the total sales of $332,450 was divided into each of the other numbers to get their percent of sales.

The total sales figure is bigger than each of the numbers it is divided into, so the answer comes out a decimal of less than 1.0 each time. Round off the answer to the third place after the decimal point, and then move the decimal point two places to the right to get percent.

For example, to find what percent the cost of goods sold is of sales for the Millbrae House Restaurant:

```
                    .4349 = .435 = 43.5%
       332,450)144,615.00
              132 980 0
               11 635 00
                9 973 50
                1 661 500
                1 329 800
                  331 7000
                  299 2050
                   32 4950
```

Millbrae House Restaurant
Income Statement
For Year Ending December 31, 19—

Income:			
Sales		$332,450	100%
Cost of Food & Beverages Sold:			
Food & Bev. Inventory, Jan. 1	$17,692		
Purchases	145,898		
Total Food & Bev. Available for Sale	163,590		
Less Food & Bev. Inventory, Dec. 31	-18,975		
Cost of Food & Bev. Sold	144,615	-144,615	43.5
Gross Profit		187,835	56.5
Expenses:			
Salaries	86,440		26.0
Payroll Taxes	8,652		2.6
Employee Welfare & Pensions	13,830		4.2
Taxes & Licenses	665		.2
Utilities	1,427		.4
Rent	12,000		3.6
Insurance	1,329		.4
Laundry	2,327		.7
Repairs & Maintenance	1,662		.5
Depreciation	3,325		1.0
Supplies	665		.2
Telephone	997		.3
Loan Interest	1,330		.4
Legal & Accounting	1,325		.4
Advertising	2,000		.6
Miscellaneous	1,650		.5
Total Expenses	139,624	-139,624	42.0
Net Income or ⟨Loss⟩		$48,211	14.5

Note: In the illustration above and in the problems in this section, no cents amounts are used. All figures have been rounded off to even dollars. Businessmen frequently do this in making up financial statements to make them easier to read.

PROBLEMS

Use the income statements you made out for the previous set of problems. Figure the percent of sales for each item on the income statement that you made up for the San Mateo Buffet in problem No. 1 in the last lesson. Write the percentage figures along the right edge of the statement as shown in the illustration for the Millbrae House Restaurant.

Here are two tips by which you can tell if you have done your figuring of percents correctly:

1. The percent for Cost of Food and Beverage Sold plus the percent for Gross Profit will total 100%.

2. The percents for cost of Food and Beverage Sold plus Total Expenses plus Net Income will total approximately 100%. (They may total 1 or 2 percent above or below 100% because you have rounded off some numbers.)

Figure the percent of sales on each of the other five income statements you made out for the last lesson.

QUESTIONS

1. What are two things financial statements are used for?
2. Briefly, what does a balance sheet show?
3. How do we find net worth?
4. In business vocabulary, what do the following terms mean?
    a. Cash
    b. Accounts receivable
    c. Accounts payable
    d. Capital
    e. Assets
    f. Liabilities
    g. Gross profit
    h. Net profit
5. Briefly, what is an income statement?
6. Where does one get the information for an income statement?
7. Why do businesspeople figure percentages of sales?

DEPRECIATION

One of the items of expense found in statements of income and expense is depreciation. Depreciation means the wearing out of something. Depreciation expense is the loss in value of something that is wearing out.

If you buy a car for $6,000 and drive it for a year, it is worth less than $6,000 at the end of the year. It has worn out a certain amount. It has less value than it had when you purchased it. If the $6,000 car is worth only $4,500 at the end of the year, we say it has depreciated $1,500. In order to figure profit or loss accurately, the businessman must estimate accurately how much his assets are going to depreciate.

Government Regulations

The government has rules regarding depreciation for businesspeople reporting their profits for income taxes. The purpose of the government's rules on depreciation is to make certain that depreciation is estimated honestly and profits are figured honestly.

Methods of Depreciation

There are a number of methods of figuring depreciation that the government allows for tax purposes. Three well-known methods are: straight-line method, declining-balance method, and sum-of-the-digits method.

Many businesspeople use the straight-line method of figuring depreciation because it is easy to figure and it is quite accurate.

Straight-line Method

Straight-line depreciation simply means dividing the cost of an asset by the number of years it will probably be useful. The answer is the depreciation expense for one year. For example, if a piece of furniture costs $150 and it is estimated that the furniture will be useful for 10 years, then it depreciates in value 1/10, or $15, each year until it is worth nothing.

Salvage Value

The asset, such as furniture, might be expected to have some small value left as junk or for trade-in on new assets at the end of its useful life. This estimated value when an asset's useful life ends is known as salvage value or scrap value.

If it is expected that an asset will have some salvage value, the salvage amount is deducted from the cost of the asset before dividing by the number of years of useful life.

How to Figure Depreciation

Here is an example of depreciation expense figured on the straight-line method for a new refrigerator:

```
Cost of refrigerator = $400
Less salvage value    −50
                     $350
```

Useful life expected = 10 years
Depreciation expense for one year = 1/10 of $350
Depreciation expense = $35

Book Value

At the end of the first year the $400 refrigerator is estimated to be worth $365. This estimated value, the cost minus the depreciation, is called the book value of an asset.

PROBLEMS

In each of the following problems, find: (a) the amount on which depreciation will be figured (cost minus salvage value), (b) the amount of depreciation to be taken each year, and (c) the book value after one year.

1. Twenty television sets, total cost $5,000, total salvage value $500, four-year life.

2. Furniture, total cost $23,000, total salvage value $2,300, ten-year life.

3. Carpets, total cost $32,000, no salvage value, six-year life.

4. Six water heaters, total cost $2,700, total salvage value $360, three-year life.

5. Drapes, total cost $2,250, no salvage value, five-year life.

6. Large kitchen equipment, total cost $31,750, total salvage value $3,175, fifteen-year life.

7. Two Ford station wagons, total cost $14,400, total salvage value $2,400, three-year life.

8. Five vacuum cleaners, total cost $725, total salvage value $75, four-year life.

9. Five cleaning carts, total cost $415, total salvage value $50, ten-year life.

10. Twenty motel units, built of concrete, total cost $382,500, no salvage value, forty-year life.

QUESTIONS

1. What is depreciation?
2. What is the reason for government income tax rules on depreciation?
3. How does straight-line depreciation work?
4. What is salvage value?
5. What is book value?

LOAN PAYMENTS AND INTEREST

Hotels, motels, and restaurants have to spend a great deal of money to get started in business. For example, it can cost $40,000 a unit or more to build a concrete motel. That is, if you build a modern motel with fifty units, it will cost about $2,000,000. Then you have to buy furniture, carpets, drapes, linens, water heaters, and so on.

How does a motel owner pay for all these things? Even if the new motel is being built for a large company that has been in business for many years, the company may not have the cash to pay out before it starts taking in any money from the new motel. Therefore, money has to be borrowed or some pieces of furniture and equipment have to be purchased on credit.

When a businessperson or a consumer borrows money or buys merchandise and pays for it in installments, he or she must pay interest. Rates of interest have changed a number of times in the last few years. They have been as low as 8 or 9 percent and as high as 20 percent or more. The rate depends on business conditions at the time of the loan, who is lending the money to whom, and for what purpose.

Where to Borrow

The borrower shops around at banks to find the lowest interest rate. He or she finds out how much interest would be owed if the seller were paid in installments as compared to borrowing from a bank. Then he or she signs an agreement for the best deal.

How to Repay

Usually, bank loans or large credit purchases from furniture and equipment companies are paid back in equal monthly payments. Part of the monthly payment pays the interest on the loan for one month. The remainder of the payment is a payment on the loan itself, which is called the "principal."

Amount of Interest Changes

Even though the total payment is the same each month, the amount of interest changes monthly and the amount used to pay off the principal changes. As each payment is made and the principal of the loan gets smaller, the amount of interest due each month is smaller. This leaves a larger part of the monthly payment to be applied to the principal.

Pictured on the next page is a simple loan chart which shows how the amount of interest and the amount paid on the principal change monthly.

Amount of original loan = $1,000   Monthly payments = $100

Rate of interest = 9% per year (1/12 of 9% per month)

Date	Total Monthly Payment	Amount of Interest Paid (for 1 month)	Amount Paid on Principal	Balance Due on Principal
Jan. 1				$1,000.00
Feb. 1	$100.00	$7.50	$92.50	907.50
Mar. 1	100.00	6.81	93.19	814.31
Apr. 1	100.00	6.11	93.89	720.42

In the illustration above, the figuring for the February 1 payment was done as follows:

Amount of Interest Paid

.09 × $1,000.00 = $90.00
$90.00 ÷ 12 = $7.50

Amount Paid on Principal

$100.00 − $7.50 = $92.50

Balance Due on Principal

$1,000.00 − $92.50 = $907.50

For the March 1 payment, the interest was figured on the principal that was owed for the month of February, $907.50.

PROBLEMS

Make loan charts similar to the illustration for each of the following loans. Show the payments on each loan for 5 months.

1. June 1, amount of original loan, $5,000.00
    monthly payments, $150.00
    rate of interest, 12% per year

2. March 15, amount of original loan, $3,500.00
    monthly payments, $135.00
    rate of interest, 9% per year

3. February 1, amount of original loan, $2,500.00
    monthly payments, $200.00
    rate of interest, 8½% per year

4. July 1, amount of original loan, $10,000.00
    monthly payments, $125.00
    rate of interest, 9½% per year

5. May 1, amount of original loan, $4,000.00
    monthly payments, $115.00
    rate of interest, 8¾% per year

QUESTIONS

1. Approximately how much might it cost to build a concrete motel?

2. Where do businesspeople get money to build a motel if they don't have the cash?

3. About how much might interest rates be on bank loans and purchases on credit?

4. Explain how the amounts paid on interest and principal change each month, although the monthly installments are equal.

INCOME TAXES—"PAY AS YOU GO"*

Young people going to work for the first time are usually shocked to find how much money is taken out of their paychecks for various reasons, mostly taxes. The two major taxes are Federal income taxes and social security taxes. Social security taxes are technically known as payments under the Federal Insurance Contributions Act, or F.I.C.A.

Both of these taxes are "pay as you go," or as you earn the money during the year. Employers are required by law to withhold income tax payments and social security tax payments from your paycheck. The employer contributes to your social security an amount equal to the amount deducted from your paycheck. He or she sends the money withheld, plus the employer's social security contribution, to the government for your income tax account and your social security account.

At the end of the year, when you file your income tax return, the total taxes due on your earnings should be very close to the amount that has been withheld from your paychecks. If not enough was withheld, you must send the Internal Revenue Service a check for the difference. If too much money was withheld, the Internal Revenue Service will send you a refund.

In some cases, it is possible that too much social security tax has been withheld and sent to the government for your account. When you file your tax return, there is a place to show this overpayment; and the government will refund the social security overpayment to you.

ESTIMATED TAX PAYMENTS

If you are self-employed and not receiving a paycheck from anyone, you are required to send to the I.R.S. four times a year, payments based on what you estimate your taxes for the year to be.

You might be receiving a paycheck from an employer and, in addition, have some other income such as interest, or dividends, or rent. If the amount being withheld will not be enough to equal your total income tax due for the year because of this extra income, you will have to make estimated tax payments in addition to your withholding. One way or the other, the government collects the taxes as you earn the money.

---

* The information in this section is based on the tax laws in effect for 1987. Some details of the laws are changed each year, but the principles involved remain the same, including the use of withholding tables as used in the problem at the end of this lesson.

PENALTIES

There are penalties for failing to pay enough taxes through withholding and estimates "as you go" during the year. The law allows you to underestimate by a small percentage; but if your underpayment is more than the percentage allowed, you will be charged the penalty.

There are also penalties for:

a) Failing to file a tax return if you are legally required to do so.

b) Filing the return late.

c) Paying your taxes late.

d) Not paying the taxes you owe.

e) Not reporting income you have that is taxable.

The penalties are usually fines; but in serious cases where people willfully evade paying their proper taxes, they can be given a prison sentence.

WHO MUST FILE A FEDERAL INCOME TAX RETURN?

Every citizen or resident of the United States who earns a certain amount of income must file an income tax return with the Internal Revenue Service at the end of the year. If your total taxable income is less than the minimum amount set by the government, you do not have to file a return or pay taxes.

The tax laws are passed by the U.S. Congress, and they apply to everyone, regardless of age, although there are different rules that apply to people of different age groups. If a small child has taxable income that exceeds the minimum exempt figure for dependent children, a tax return must be filed for that child. The details of the tax laws and some of the figures involved change from year to year. It is the duty of every citizen to find out what the requirements are each year to determine whether or not he/she must file a tax return and pay taxes.

WHAT KIND OF INCOME IS TAXABLE?

Basically anything of value that is earned is considered income for tax purposes. In addition to salaries and wages, this includes tips, commissions, bonuses, prizes given to workers for outstanding performance on the job, interest received, dividends, rents, royalties, profits from a business, etc. If the worker's prize is given in the form of a car or television set, for example, the value of that prize would have to be reported as income for taxes.

The Internal Revenue Code lists things that are not considered income, such as gifts that were not earned, welfare benefits, certain insurance payments, etc. In general, any income is taxable unless it is specifically mentioned in the laws as not being taxable.

There are regulations that apply especially to people in certain industries.

TAXES ON GRATUITIES, TIPS

Waiters and waitresses, buspersons, cab drivers, and other people who receive tips must pay income taxes and social security taxes on the tips.

The government has ruled that tips are "earned," not given to you. The I.R.S. regulations list the following main points regarding tips:

1. All tips are income for tax purposes. You should keep a daily record of them. If the tips for a month are less than $20, you simply report that amount with your other income on your tax return at the end of the year.

2. If the tips in one month are $20 or more, you must report the amount of tips in writing to your employer (who usually has a form for you to fill out). This report must be made within ten days of the end of the month in which the tips were received.

3. An *employer must* deduct the proper withholding taxes for these tips out of your salary, or you must give the employer enough money out of your tips to cover withholding taxes on the tips.

4. The *employer must* also deduct social security taxes for the tips out of your salary or collect this tax from you. The employer does not have to make social security contributions to the employee's account on the tips, however.

5. *You, the employee, face the possibility of severe penalties from the government if you willfully neglect to report your tip income.*

TAXES ON FREE MEALS AND LODGING

What about meals that waiters, waitresses, cooks, cashiers, and other people working in restaurants eat but do not pay for? And how about the rooms that the manager of a hotel and his family live in at the hotel free of charge? Do these people have to pay taxes on the value of these meals and rooms given to them while they are at work?

No, usually they do not. There is another regulation among the tax rules that says that meals and lodging given to employees at the place where they are working and given "for the convenience of the employer" are not taxable.

Even though the employee receives the benefits, it is usually for the convenience of the employer that the cook, waitress, or manager eats at the restaurant or sleeps in a room in the hotel. In the case of lodging, you must accept the lodging at your employer's place of business as a condition of employment in order for it to be nontaxable as income.

At the end of the year the employer makes out a W-2 statement which shows how much the employee earned and how much was withheld from his or her wages for income taxes and social security taxes. One copy goes to the government and two copies to the employee. On this W-2 statement the value of the meals is usually included in the total wages paid. Then it is shown as a separate amount marked "meals," so that the government and the taxpayer know how much to deduct from the total wages in order to avoid paying income taxes on the value of the meals.

SOCIAL SECURITY NUMBERS

Every employee must have a social security number. If you do not have one, you should file a form to get one with the Social Security Administration nearest you. From the first day you start working in most occupations, you and your employer start making social security tax payments which are recorded in your social security file and determine how much you will collect in social security benefits from the government when you reach retirement age. You need a social security number along with your name when you file your income tax return. It is also useful for identification in business situations, such as cashing checks.

When children reach the age of 5, they must have a social security number even if they do not have income for tax purposes. The social security number must be shown with the child's name when the child is listed as a dependent on the parents' or guardians' income tax return.

WHERE TO GET HELP ON INCOME TAXES

There are many rules and regulations in the tax laws, and they are changed to some extent every year. Most people need expert help in following the laws. It begins with the form the employee must file with the employer when he is first hired. This is called a W-4 form, and it must be filled out properly so that the employer will make the proper withholding deduction from the employee's paycheck. It is equally important that the taxpayer understand and fill out correctly the estimated tax form if one is needed, and finally, the income tax return at the end of the year.

You can always get information and help on your federal (United States Government) taxes from the Internal Revenue Service office nearest you. There is one located in every large city of the United States and in many small towns. You can find it by looking in the telephone book under "United States Government, Internal Revenue Service." They will answer questions for you over the telephone. If you go to the I.R.S. office, someone will help you answer questions; and it will cost you nothing.

You can also get help in preparing your tax returns from public accountants. These people are licensed by the state in which they work, and their business is to do accounting for people and to help them with their taxes for a fee. The accountant signs the returns because he or she is responsible for making them out correctly. The taxpayer also signs the returns, and he/she is responsible for giving the accountant true and accurate information.

To help you with your income taxes, an accountant would charge anywhere from $25 to $100 or more, depending on how complicated your tax situation was. Frequently, a public accountant, or tax consultant, can save you much more money than the fee you pay because he or she knows how to make out your tax returns properly so that you pay as little taxes as you are legally required to pay.

STATE INCOME TAXES

State income tax laws are a little different in each state. In some states, such as Nevada, there is no state income tax at all. In some states, such as California, the rules are similar to the rules for the federal income taxes. The rate, and amount, of taxes you pay to a state government is lower than the federal rate. Your employer might make withholding deductions from your paycheck for state income taxes along with your federal income tax withholding and social security taxes.

QUESTIONS

1. What are the two major taxes employers deduct from employees' paychecks?
2. Which of these two taxes does the employer contribute to?
3. If too much income tax or social security tax is taken out of your paycheck, will you get it back?
4. If you are self-employed, how do you pay your taxes as you go?
5. What happens if you do not pay enough taxes as you go during the year?
6. Who must file an income tax return?
7. Who makes the tax laws?
8. Are children exempt from the tax laws?
9. What is the basic rule that determines if income is taxable?
10. Why are tips taxable?
11. If you earned $50 a month in tips, would you have to pay taxes on the tips as you go during the year?
12. If you are working in a restaurant and get free meals while you are on the job, do you have to pay taxes on the value of the meals?
13. Does a 6-year old child need a social security number if he or she has no income?
14. Where can you get free help with your income taxes?
15. Do all states collect income taxes?

PAYROLL RECORD

Employers are required by law to keep payroll records. They must keep a record of the total earnings of their employees, the amount of tips reported by them, the amount of withholding taxes kept out of the employees' paychecks for the government, the amount of social security taxes withheld from their checks, and in some states other deductions, such as unemployment and disability insurance.

On the following pages are: (a) a payroll record to be completed, (b) pages from the Federal Income Tax Withholding Tables for Weekly Payroll Periods for single persons and for married persons,* and (c) pages from the Social Security (FICA) Employee Tax Table.

* There are other methods permitted by the I.R.S. for figuring income tax withholding for employees. One of these is the percentage method. The I.R.S. provides tables for the employers if the employee and the employer choose to use that method. The basic idea of withholding, to "pay as you go," is the same no matter which withholding method is used.

PAYROLL RECORD

No.	Employee's Name	Single or Married	No. of Exemptions	Total Wages	Tips Reported	Total Wages Plus Tips	Withholding Tax on Wages Plus Tips	Social Sec. Tax on Wages Plus Tips	Total Deductions	Net Pay Total Wages Less Deductions
1	Bruce, Katherine	M	4	$244.00	$62.25	$306.25	$16.00	$21.90	$37.90	$206.10
2	Farmer, Elaine	S	1	222.00	81.45	303.45	37.00	21.70	58.70	163.30
3	Fry, Jean	M	2	218.00	91.35					
4	Grimes, Bob	M	4	218.00	93.40					
5	Hall, Lois	M	3	210.00	99.75					
6	Henry, Marjorie	S	1	211.50	80.05					
7	Jermyn, Ed	M	2	221.50	102.50					
8	Jermyn, James	M	5	219.00	97.75					
9	Johnson, Jim	M	6	222.00	118.00					
10	Kearns, Larry	M	3	325.00	—0—					
11	Laclergue, Don	S	1	206.25	101.10					
12	McDowell, Ron	M	2	210.00	106.90					
13	Savinar, Emily	S	1	112.00	87.00					
14	Skellenger, Jon	S	1	212.00	101.30					
15	Tegnell, John	M	2	216.00	114.00					
16	Tjader, Bev	M	4	220.00	—0—					
17	Vincent, Chris	M	3	214.00	118.60					
18	Williams, George	M	3	219.00	91.20					
19	Young, Gordy	S	1	203.50	79.20					
20	Ziegler, Harry	M	5	222.00	107.50					

Permission is hereby granted to the teacher using this book to reproduce this page as needed.

SINGLE Persons—WEEKLY Payroll Period
(For Wages Paid After December 1986)

Federal Income Tax Withholding Table

And the wages are—		And the number of withholding allowances claimed is—										
At least	But less than	0	1	2	3	4	5	6	7	8	9	10
		The amount of income tax to be withheld shall be—										
$0	$16	$0	$0	$0	$0	$0	$0	$0	$0	$0	$0	$0
16	18	1	0	0	0	0	0	0	0	0	0	0
18	20	1	0	0	0	0	0	0	0	0	0	0
20	22	1	0	0	0	0	0	0	0	0	0	0
22	24	1	0	0	0	0	0	0	0	0	0	0
24	26	1	0	0	0	0	0	0	0	0	0	0
26	28	2	0	0	0	0	0	0	0	0	0	0
28	30	2	0	0	0	0	0	0	0	0	0	0
30	32	2	0	0	0	0	0	0	0	0	0	0
32	34	2	0	0	0	0	0	0	0	0	0	0
34	36	2	0	0	0	0	0	0	0	0	0	0
36	38	3	0	0	0	0	0	0	0	0	0	0
38	40	3	0	0	0	0	0	0	0	0	0	0
40	42	3	0	0	0	0	0	0	0	0	0	0
42	44	3	0	0	0	0	0	0	0	0	0	0
44	46	4	0	0	0	0	0	0	0	0	0	0
46	48	4	0	0	0	0	0	0	0	0	0	0
48	50	4	0	0	0	0	0	0	0	0	0	0
50	52	4	0	0	0	0	0	0	0	0	0	0
52	54	5	0	0	0	0	0	0	0	0	0	0
54	56	5	1	0	0	0	0	0	0	0	0	0
56	58	5	1	0	0	0	0	0	0	0	0	0
58	60	6	1	0	0	0	0	0	0	0	0	0
60	62	6	1	0	0	0	0	0	0	0	0	0
62	64	6	2	0	0	0	0	0	0	0	0	0
64	66	7	2	0	0	0	0	0	0	0	0	0
66	68	7	2	0	0	0	0	0	0	0	0	0
68	70	7	2	0	0	0	0	0	0	0	0	0
70	72	7	2	0	0	0	0	0	0	0	0	0
72	74	8	3	0	0	0	0	0	0	0	0	0
74	76	8	3	0	0	0	0	0	0	0	0	0
76	78	8	3	0	0	0	0	0	0	0	0	0
78	80	9	3	0	0	0	0	0	0	0	0	0
80	82	9	4	0	0	0	0	0	0	0	0	0
82	84	9	4	0	0	0	0	0	0	0	0	0
84	86	10	4	0	0	0	0	0	0	0	0	0
86	88	10	4	0	0	0	0	0	0	0	0	0
88	90	10	5	0	0	0	0	0	0	0	0	0
90	92	10	5	1	0	0	0	0	0	0	0	0
92	94	11	5	1	0	0	0	0	0	0	0	0
94	96	11	6	1	0	0	0	0	0	0	0	0
96	98	11	6	1	0	0	0	0	0	0	0	0
98	100	12	6	1	0	0	0	0	0	0	0	0
100	105	12	7	2	0	0	0	0	0	0	0	0
105	110	13	7	2	0	0	0	0	0	0	0	0
110	115	14	8	3	0	0	0	0	0	0	0	0
115	120	14	9	4	0	0	0	0	0	0	0	0
120	125	15	10	4	0	0	0	0	0	0	0	0
125	130	16	10	5	1	0	0	0	0	0	0	0
130	135	17	11	6	1	0	0	0	0	0	0	0
135	140	17	12	6	2	0	0	0	0	0	0	0
140	145	18	13	7	2	0	0	0	0	0	0	0
145	150	19	13	8	3	0	0	0	0	0	0	0
150	160	20	15	9	4	0	0	0	0	0	0	0
160	170	22	16	11	5	1	0	0	0	0	0	0
170	180	23	18	12	7	2	0	0	0	0	0	0
180	190	25	19	14	8	3	0	0	0	0	0	0
190	200	26	21	15	10	4	0	0	0	0	0	0
200	210	28	22	17	11	6	1	0	0	0	0	0
210	220	29	24	18	13	7	2	0	0	0	0	0
220	230	31	25	20	14	9	3	0	0	0	0	0
230	240	32	27	21	16	10	5	0	0	0	0	0
240	250	34	28	23	17	12	6	1	0	0	0	0
250	260	35	30	24	19	13	8	3	0	0	0	0
260	270	37	31	26	20	15	9	4	0	0	0	0
270	280	38	33	27	22	16	11	5	1	0	0	0
280	290	40	34	29	23	18	12	7	2	0	0	0
290	300	41	36	30	25	19	14	8	3	0	0	0
300	310	43	37	32	26	21	15	10	4	0	0	0
310	320	44	39	33	28	22	17	11	6	1	0	0

MARRIED Persons–**WEEKLY** Payroll Period
(For Wages Paid After December 1986)

Federal Income Tax Withholding Table

And the wages are–		And the number of withholding allowances claimed is–										
At least	But less than	0	1	2	3	4	5	6	7	8	9	10
		The amount of income tax to be withheld shall be–										
$0	$40	$0	$0	$0	$0	$0	$0	$0	$0	$0	$0	$0
40	42	1	0	0	0	0	0	0	0	0	0	0
42	44	1	0	0	0	0	0	0	0	0	0	0
44	46	1	0	0	0	0	0	0	0	0	0	0
46	48	1	0	0	0	0	0	0	0	0	0	0
48	50	1	0	0	0	0	0	0	0	0	0	0
50	52	2	0	0	0	0	0	0	0	0	0	0
52	54	2	0	0	0	0	0	0	0	0	0	0
54	56	2	0	0	0	0	0	0	0	0	0	0
56	58	2	0	0	0	0	0	0	0	0	0	0
58	60	3	0	0	0	0	0	0	0	0	0	0
60	62	3	0	0	0	0	0	0	0	0	0	0
62	64	3	0	0	0	0	0	0	0	0	0	0
64	66	3	0	0	0	0	0	0	0	0	0	0
66	68	3	0	0	0	0	0	0	0	0	0	0
68	70	4	0	0	0	0	0	0	0	0	0	0
70	72	4	0	0	0	0	0	0	0	0	0	0
72	74	4	0	0	0	0	0	0	0	0	0	0
74	76	4	0	0	0	0	0	0	0	0	0	0
76	78	5	1	0	0	0	0	0	0	0	0	0
78	80	5	1	0	0	0	0	0	0	0	0	0
80	82	5	1	0	0	0	0	0	0	0	0	0
82	84	5	1	0	0	0	0	0	0	0	0	0
84	86	5	1	0	0	0	0	0	0	0	0	0
86	88	6	2	0	0	0	0	0	0	0	0	0
88	90	6	2	0	0	0	0	0	0	0	0	0
90	92	6	2	0	0	0	0	0	0	0	0	0
92	94	6	2	0	0	0	0	0	0	0	0	0
94	96	7	2	0	0	0	0	0	0	0	0	0
96	98	7	3	0	0	0	0	0	0	0	0	0
98	100	7	3	0	0	0	0	0	0	0	0	0
100	105	8	3	0	0	0	0	0	0	0	0	0
105	110	8	4	0	0	0	0	0	0	0	0	0
110	115	9	4	0	0	0	0	0	0	0	0	0
115	120	10	5	1	0	0	0	0	0	0	0	0
120	125	11	6	2	0	0	0	0	0	0	0	0
125	130	11	6	2	0	0	0	0	0	0	0	0
130	135	12	7	3	0	0	0	0	0	0	0	0
135	140	13	7	3	0	0	0	0	0	0	0	0
140	145	14	8	4	0	0	0	0	0	0	0	0
145	150	14	9	4	0	0	0	0	0	0	0	0
150	160	16	10	5	1	0	0	0	0	0	0	0
160	170	17	12	6	2	0	0	0	0	0	0	0
170	180	19	13	8	3	0	0	0	0	0	0	0
180	190	20	15	9	4	0	0	0	0	0	0	0
190	200	22	16	11	5	1	0	0	0	0	0	0
200	210	23	18	12	7	3	0	0	0	0	0	0
210	220	25	19	14	8	4	0	0	0	0	0	0
220	230	26	21	15	10	5	1	0	0	0	0	0
230	240	28	22	17	11	6	2	0	0	0	0	0
240	250	29	24	18	13	7	3	0	0	0	0	0
250	260	31	25	20	14	9	4	0	0	0	0	0
260	270	32	27	21	16	10	5	1	0	0	0	0
270	280	34	28	23	17	12	6	2	0	0	0	0
280	290	35	30	24	19	13	8	3	0	0	0	0
290	300	37	31	26	20	15	9	4	0	0	0	0
300	310	38	33	27	22	16	11	6	1	0	0	0
310	320	40	34	29	23	18	12	7	3	0	0	0
320	330	41	36	30	25	19	14	8	4	0	0	0
330	340	43	37	32	26	21	15	10	5	1	0	0
340	350	44	39	33	28	22	17	11	6	2	0	0
350	360	46	40	35	29	24	18	13	7	3	0	0
360	370	47	42	36	31	25	20	14	9	4	0	0
370	380	49	43	38	32	27	21	16	10	5	1	0
380	390	50	45	39	34	28	23	17	12	6	2	0
390	400	52	46	41	35	30	24	19	13	8	3	0
400	410	53	48	42	37	31	26	20	15	9	4	0
410	420	55	49	44	38	33	27	22	16	11	6	2
420	430	56	51	45	40	34	29	23	18	12	7	3
430	440	58	52	47	41	36	30	25	19	14	8	4

Social Security Employee Tax Table for 1987
7.15% employee tax deductions

Wages at least	But less than	Tax to be withheld	Wages at least	But less than	Tax to be withheld	Wages at least	But less than	Tax to be withheld	Wages at least	But less than	Tax to be withheld
$0.00	$0.07	$0.00	12.66	12.80	.91	25.39	25.53	1.82	38.12	38.26	2.73
.07	.21	.01	12.80	12.94	.92	25.53	25.67	1.83	38.26	38.40	2.74
.21	.35	.02	12.94	13.08	.93	25.67	25.81	1.84	38.40	38.54	2.75
.35	.49	.03	13.08	13.22	.94	25.81	25.95	1.85	38.54	38.68	2.76
.49	.63	.04	13.22	13.36	.95	25.95	26.09	1.86	38.68	38.82	2.77
.63	.77	.05	13.36	13.50	.96	26.09	26.23	1.87	38.82	38.96	2.78
.77	.91	.06	13.50	13.64	.97	26.23	26.37	1.88	38.96	39.10	2.79
.91	1.05	.07	13.64	13.78	.98	26.37	26.51	1.89	39.10	39.24	2.80
1.05	1.19	.08	13.78	13.92	.99	26.51	26.65	1.90	39.24	39.38	2.81
1.19	1.33	.09	13.92	14.06	1.00	26.65	26.79	1.91	39.38	39.52	2.82
1.33	1.47	.10	14.06	14.20	1.01	26.79	26.93	1.92	39.52	39.66	2.83
1.47	1.61	.11	14.20	14.34	1.02	26.93	27.07	1.93	39.66	39.80	2.84
1.61	1.75	.12	14.34	14.48	1.03	27.07	27.21	1.94	39.80	39.94	2.85
1.75	1.89	.13	14.48	14.62	1.04	27.21	27.35	1.95	39.94	40.07	2.86
1.89	2.03	.14	14.62	14.76	1.05	27.35	27.49	1.96	40.07	40.21	2.87
2.03	2.17	.15	14.76	14.90	1.06	27.49	27.63	1.97	40.21	40.35	2.88
2.17	2.31	.16	14.90	15.04	1.07	27.63	27.77	1.98	40.35	40.49	2.89
2.31	2.45	.17	15.04	15.18	1.08	27.77	27.91	1.99	40.49	40.63	2.90
2.45	2.59	.18	15.18	15.32	1.09	27.91	28.05	2.00	40.63	40.77	2.91
2.59	2.73	.19	15.32	15.46	1.10	28.05	28.19	2.01	40.77	40.91	2.92
2.73	2.87	.20	15.46	15.60	1.11	28.19	28.33	2.02	40.91	41.05	2.93
2.87	3.01	.21	15.60	15.74	1.12	28.33	28.47	2.03	41.05	41.19	2.94
3.01	3.15	.22	15.74	15.88	1.13	28.47	28.61	2.04	41.19	41.33	2.95
3.15	3.29	.23	15.88	16.02	1.14	28.61	28.75	2.05	41.33	41.47	2.96
3.29	3.43	.24	16.02	16.16	1.15	28.75	28.89	2.06	41.47	41.61	2.97
3.43	3.57	.25	16.16	16.30	1.16	28.89	29.03	2.07	41.61	41.75	2.98
3.57	3.71	.26	16.30	16.44	1.17	29.03	29.17	2.08	41.75	41.89	2.99
3.71	3.85	.27	16.44	16.58	1.18	29.17	29.31	2.09	41.89	42.03	3.00
3.85	3.99	.28	16.58	16.72	1.19	29.31	29.45	2.10	42.03	42.17	3.01
3.99	4.13	.29	16.72	16.86	1.20	29.45	29.59	2.11	42.17	42.31	3.02
4.13	4.27	.30	16.86	17.00	1.21	29.59	29.73	2.12	42.31	42.45	3.03
4.27	4.41	.31	17.00	17.14	1.22	29.73	29.87	2.13	42.45	42.59	3.04
4.41	4.55	.32	17.14	17.28	1.23	29.87	30.00	2.14	42.59	42.73	3.05
4.55	4.69	.33	17.28	17.42	1.24	30.00	30.14	2.15	42.73	42.87	3.06
4.69	4.83	.34	17.42	17.56	1.25	30.14	30.28	2.16	42.87	43.01	3.07
4.83	4.97	.35	17.56	17.70	1.26	30.28	30.42	2.17	43.01	43.15	3.08
4.97	5.11	.36	17.70	17.84	1.27	30.42	30.56	2.18	43.15	43.29	3.09
5.11	5.25	.37	17.84	17.98	1.28	30.56	30.70	2.19	43.29	43.43	3.10
5.25	5.39	.38	17.98	18.12	1.29	30.70	30.84	2.20	43.43	43.57	3.11
5.39	5.53	.39	18.12	18.26	1.30	30.84	30.98	2.21	43.57	43.71	3.12
5.53	5.67	.40	18.26	18.40	1.31	30.98	31.12	2.22	43.71	43.85	3.13
5.67	5.81	.41	18.40	18.54	1.32	31.12	31.26	2.23	43.85	43.99	3.14
5.81	5.95	.42	18.54	18.68	1.33	31.26	31.40	2.24	43.99	44.13	3.15
5.95	6.09	.43	18.68	18.82	1.34	31.40	31.54	2.25	44.13	44.27	3.16
6.09	6.23	.44	18.82	18.96	1.35	31.54	31.68	2.26	44.27	44.41	3.17
6.23	6.37	.45	18.96	19.10	1.36	31.68	31.82	2.27	44.41	44.55	3.18
6.37	6.51	.46	19.10	19.24	1.37	31.82	31.96	2.28	44.55	44.69	3.19
6.51	6.65	.47	19.24	19.38	1.38	31.96	32.10	2.29	44.69	44.83	3.20
6.65	6.79	.48	19.38	19.52	1.39	32.10	32.24	2.30	44.83	44.97	3.21
6.79	6.93	.49	19.52	19.66	1.40	32.24	32.38	2.31	44.97	45.11	3.22
6.93	7.07	.50	19.66	19.80	1.41	32.38	32.52	2.32	45.11	45.25	3.23
7.07	7.21	.51	19.80	19.94	1.42	32.52	32.66	2.33	45.25	45.39	3.24
7.21	7.35	.52	19.94	20.07	1.43	32.66	32.80	2.34	45.39	45.53	3.25
7.35	7.49	.53	20.07	20.21	1.44	32.80	32.94	2.35	45.53	45.67	3.26
7.49	7.63	.54	20.21	20.35	1.45	32.94	33.08	2.36	45.67	45.81	3.27
7.63	7.77	.55	20.35	20.49	1.46	33.08	33.22	2.37	45.81	45.95	3.28
7.77	7.91	.56	20.49	20.63	1.47	33.22	33.36	2.38	45.95	46.09	3.29
7.91	8.05	.57	20.63	20.77	1.48	33.36	33.50	2.39	46.09	46.23	3.30
8.05	8.19	.58	20.77	20.91	1.49	33.50	33.64	2.40	46.23	46.37	3.31
8.19	8.33	.59	20.91	21.05	1.50	33.64	33.78	2.41	46.37	46.51	3.32
8.33	8.47	.60	21.05	21.19	1.51	33.78	33.92	2.42	46.51	46.65	3.33
8.47	8.61	.61	21.19	21.33	1.52	33.92	34.06	2.43	46.65	46.79	3.34
8.61	8.75	.62	21.33	21.47	1.53	34.06	34.20	2.44	46.79	46.93	3.35
8.75	8.89	.63	21.47	21.61	1.54	34.20	34.34	2.45	46.93	47.07	3.36
8.89	9.03	.64	21.61	21.75	1.55	34.34	34.48	2.46	47.07	47.21	3.37
9.03	9.17	.65	21.75	21.89	1.56	34.48	34.62	2.47	47.21	47.35	3.38
9.17	9.31	.66	21.89	22.03	1.57	34.62	34.76	2.48	47.35	47.49	3.39
9.31	9.45	.67	22.03	22.17	1.58	34.76	34.90	2.49	47.49	47.63	3.40
9.45	9.59	.68	22.17	22.31	1.59	34.90	35.04	2.50	47.63	47.77	3.41
9.59	9.73	.69	22.31	22.45	1.60	35.04	35.18	2.51	47.77	47.91	3.42
9.73	9.87	.70	22.45	22.59	1.61	35.18	35.32	2.52	47.91	48.05	3.43
9.87	10.00	.71	22.59	22.73	1.62	35.32	35.46	2.53	48.05	48.19	3.44
10.00	10.14	.72	22.73	22.87	1.63	35.46	35.60	2.54	48.19	48.33	3.45
10.14	10.28	.73	22.87	23.01	1.64	35.60	35.74	2.55	48.33	48.47	3.46
10.28	10.42	.74	23.01	23.15	1.65	35.74	35.88	2.56	48.47	48.61	3.47
10.42	10.56	.75	23.15	23.29	1.66	35.88	36.02	2.57	48.61	48.75	3.48
10.56	10.70	.76	23.29	23.43	1.67	36.02	36.16	2.58	48.75	48.89	3.49
10.70	10.84	.77	23.43	23.57	1.68	36.16	36.30	2.59	48.89	49.03	3.50
10.84	10.98	.78	23.57	23.71	1.69	36.30	36.44	2.60	49.03	49.17	3.51
10.98	11.12	.79	23.71	23.85	1.70	36.44	36.58	2.61	49.17	49.31	3.52
11.12	11.26	.80	23.85	23.99	1.71	36.58	36.72	2.62	49.31	49.45	3.53
11.26	11.40	.81	23.99	24.13	1.72	36.72	36.86	2.63	49.45	49.59	3.54
11.40	11.54	.82	24.13	24.27	1.73	36.86	37.00	2.64	49.59	49.73	3.55
11.54	11.68	.83	24.27	24.41	1.74	37.00	37.14	2.65	49.73	49.87	3.56
11.68	11.82	.84	24.41	24.55	1.75	37.14	37.28	2.66	49.87	50.00	3.57
11.82	11.96	.85	24.55	24.69	1.76	37.28	37.42	2.67	50.00	50.14	3.58
11.96	12.10	.86	24.69	24.83	1.77	37.42	37.56	2.68	50.14	50.28	3.59
12.10	12.24	.87	24.83	24.97	1.78	37.56	37.70	2.69	50.28	50.42	3.60
12.24	12.38	.88	24.97	25.11	1.79	37.70	37.84	2.70	50.42	50.56	3.61
12.38	12.52	.89	25.11	25.25	1.80	37.84	37.98	2.71	50.56	50.70	3.62
12.52	12.66	.90	25.25	25.39	1.81	37.98	38.12	2.72	50.70	50.84	3.63

Social Security Employee Tax Table for 1987
7.15% employee tax deductions

Wages at least	But less than	Tax to be withheld	Wages at least	But less than	Tax to be withheld	Wages at least	But less than	Tax to be withheld	Wages at least	But less than	Tax to be withheld
50.84	50.98	3.64	63.57	63.71	4.55	76.30	76.44	5.46	89.03	89.17	6.37
50.98	51.12	3.65	63.71	63.85	4.56	76.44	76.58	5.47	89.17	89.31	6.38
51.12	51.26	3.66	63.85	63.99	4.57	76.58	76.72	5.48	89.31	89.45	6.39
51.26	51.40	3.67	63.99	64.13	4.58	76.72	76.86	5.49	89.45	89.59	6.40
51.40	51.54	3.68	64.13	64.27	4.59	76.86	77.00	5.50	89.59	89.73	6.41
51.54	51.68	3.69	64.27	64.41	4.60	77.00	77.14	5.51	89.73	89.87	6.42
51.68	51.82	3.70	64.41	64.55	4.61	77.14	77.28	5.52	89.87	90.00	6.43
51.82	51.96	3.71	64.55	64.69	4.62	77.28	77.42	5.53	90.00	90.14	6.44
51.96	52.10	3.72	64.69	64.83	4.63	77.42	77.56	5.54	90.14	90.28	6.45
52.10	52.24	3.73	64.83	64.97	4.64	77.56	77.70	5.55	90.28	90.42	6.46
52.24	52.38	3.74	64.97	65.11	4.65	77.70	77.84	5.56	90.42	90.56	6.47
52.38	52.52	3.75	65.11	65.25	4.66	77.84	77.98	5.57	90.56	90.70	6.48
52.52	52.66	3.76	65.25	65.39	4.67	77.98	78.12	5.58	90.70	90.84	6.49
52.66	52.80	3.77	65.39	65.53	4.68	78.12	78.26	5.59	90.84	90.98	6.50
52.80	52.94	3.78	65.53	65.67	4.69	78.26	78.40	5.60	90.98	91.12	6.51
52.94	53.08	3.79	65.67	65.81	4.70	78.40	78.54	5.61	91.12	91.26	6.52
53.08	53.22	3.80	65.81	65.95	4.71	78.54	78.68	5.62	91.26	91.40	6.53
53.22	53.36	3.81	65.95	66.09	4.72	78.68	78.82	5.63	91.40	91.54	6.54
53.36	53.50	3.82	66.09	66.23	4.73	78.82	78.96	5.64	91.54	91.68	6.55
53.50	53.64	3.83	66.23	66.37	4.74	78.96	79.10	5.65	91.68	91.82	6.56
53.64	53.78	3.84	66.37	66.51	4.75	79.10	79.24	5.66	91.82	91.96	6.57
53.78	53.92	3.85	66.51	66.65	4.76	79.24	79.38	5.67	91.96	92.10	6.58
53.92	54.06	3.86	66.65	66.79	4.77	79.38	79.52	5.68	92.10	92.24	6.59
54.06	54.20	3.87	66.79	66.93	4.78	79.52	79.66	5.69	92.24	92.38	6.60
54.20	54.34	3.88	66.93	67.07	4.79	79.66	79.80	5.70	92.38	92.52	6.61
54.34	54.48	3.89	67.07	67.21	4.80	79.80	79.94	5.71	92.52	92.66	6.62
54.48	54.62	3.90	67.21	67.35	4.81	79.94	80.07	5.72	92.66	92.80	6.63
54.62	54.76	3.91	67.35	67.49	4.82	80.07	80.21	5.73	92.80	92.94	6.64
54.76	54.90	3.92	67.49	67.63	4.83	80.21	80.35	5.74	92.94	93.08	6.65
54.90	55.04	3.93	67.63	67.77	4.84	80.35	80.49	5.75	93.08	93.22	6.66
55.04	55.18	3.94	67.77	67.91	4.85	80.49	80.63	5.76	93.22	93.36	6.67
55.18	55.32	3.95	67.91	68.05	4.86	80.63	80.77	5.77	93.36	93.50	6.68
55.32	55.46	3.96	68.05	68.19	4.87	80.77	80.91	5.78	93.50	93.64	6.69
55.46	55.60	3.97	68.19	68.33	4.88	80.91	81.05	5.79	93.64	93.78	6.70
55.60	55.74	3.98	68.33	68.47	4.89	81.05	81.19	5.80	93.78	93.92	6.71
55.74	55.88	3.99	68.47	68.61	4.90	81.19	81.33	5.81	93.92	94.06	6.72
55.88	56.02	4.00	68.61	68.75	4.91	81.33	81.47	5.82	94.06	94.20	6.73
56.02	56.16	4.01	68.75	68.89	4.92	81.47	81.61	5.83	94.20	94.34	6.74
56.16	56.30	4.02	68.89	69.03	4.93	81.61	81.75	5.84	94.34	94.48	6.75
56.30	56.44	4.03	69.03	69.17	4.94	81.75	81.89	5.85	94.48	94.62	6.76
56.44	56.58	4.04	69.17	69.31	4.95	81.89	82.03	5.86	94.62	94.76	6.77
56.58	56.72	4.05	69.31	69.45	4.96	82.03	82.17	5.87	94.76	94.90	6.78
56.72	56.86	4.06	69.45	69.59	4.97	82.17	82.31	5.88	94.90	95.04	6.79
56.86	57.00	4.07	69.59	69.73	4.98	82.31	82.45	5.89	95.04	95.18	6.80
57.00	57.14	4.08	69.73	69.87	4.99	82.45	82.59	5.90	95.18	95.32	6.81
57.14	57.28	4.09	69.87	70.00	5.00	82.59	82.73	5.91	95.32	95.46	6.82
57.28	57.42	4.10	70.00	70.14	5.01	82.73	82.87	5.92	95.46	95.60	6.83
57.42	57.56	4.11	70.14	70.28	5.02	82.87	83.01	5.93	95.60	95.74	6.84
57.56	57.70	4.12	70.28	70.42	5.03	83.01	83.15	5.94	95.74	95.88	6.85
57.70	57.84	4.13	70.42	70.56	5.04	83.15	83.29	5.95	95.88	96.02	6.86
57.84	57.98	4.14	70.56	70.70	5.05	83.29	83.43	5.96	96.02	96.16	6.87
57.98	58.12	4.15	70.70	70.84	5.06	83.43	83.57	5.97	96.16	96.30	6.88
58.12	58.26	4.16	70.84	70.98	5.07	83.57	83.71	5.98	96.30	96.44	6.89
58.26	58.40	4.17	70.98	71.12	5.08	83.71	83.85	5.99	96.44	96.58	6.90
58.40	58.54	4.18	71.12	71.26	5.09	83.85	83.99	6.00	96.58	96.72	6.91
58.54	58.68	4.19	71.26	71.40	5.10	83.99	84.13	6.01	96.72	96.86	6.92
58.68	58.82	4.20	71.40	71.54	5.11	84.13	84.27	6.02	96.86	97.00	6.93
58.82	58.96	4.21	71.54	71.68	5.12	84.27	84.41	6.03	97.00	97.14	6.94
58.96	59.10	4.22	71.68	71.82	5.13	84.41	84.55	6.04	97.14	97.28	6.95
59.10	59.24	4.23	71.82	71.96	5.14	84.55	84.69	6.05	97.28	97.42	6.96
59.24	59.38	4.24	71.96	72.10	5.15	84.69	84.83	6.06	97.42	97.56	6.97
59.38	59.52	4.25	72.10	72.24	5.16	84.83	84.97	6.07	97.56	97.70	6.98
59.52	59.66	4.26	72.24	72.38	5.17	84.97	85.11	6.08	97.70	97.84	6.99
59.66	59.80	4.27	72.38	72.52	5.18	85.11	85.25	6.09	97.84	97.98	7.00
59.80	59.94	4.28	72.52	72.66	5.19	85.25	85.39	6.10	97.98	98.12	7.01
59.94	60.07	4.29	72.66	72.80	5.20	85.39	85.53	6.11	98.12	98.26	7.02
60.07	60.21	4.30	72.80	72.94	5.21	85.53	85.67	6.12	98.26	98.40	7.03
60.21	60.35	4.31	72.94	73.08	5.22	85.67	85.81	6.13	98.40	98.54	7.04
60.35	60.49	4.32	73.08	73.22	5.23	85.81	85.95	6.14	98.54	98.68	7.05
60.49	60.63	4.33	73.22	73.36	5.24	85.95	86.09	6.15	98.68	98.82	7.06
60.63	60.77	4.34	73.36	73.50	5.25	86.09	86.23	6.16	98.82	98.96	7.07
60.77	60.91	4.35	73.50	73.64	5.26	86.23	86.37	6.17	98.96	99.10	7.08
60.91	61.05	4.36	73.64	73.78	5.27	86.37	86.51	6.18	99.10	99.24	7.09
61.05	61.19	4.37	73.78	73.92	5.28	86.51	86.65	6.19	99.24	99.38	7.10
61.19	61.33	4.38	73.92	74.06	5.29	86.65	86.79	6.20	99.38	99.52	7.11
61.33	61.47	4.39	74.06	74.20	5.30	86.79	86.93	6.21	99.52	99.66	7.12
61.47	61.61	4.40	74.20	74.34	5.31	86.93	87.07	6.22	99.66	99.80	7.13
61.61	61.75	4.41	74.34	74.48	5.32	87.07	87.21	6.23	99.80	99.94	7.14
61.75	61.89	4.42	74.48	74.62	5.33	87.21	87.35	6.24	99.94	100.00	7.15
61.89	62.03	4.43	74.62	74.76	5.34	87.35	87.49	6.25			
62.03	62.17	4.44	74.76	74.90	5.35	87.49	87.63	6.26			
62.17	62.31	4.45	74.90	75.04	5.36	87.63	87.77	6.27			
62.31	62.45	4.46	75.04	75.18	5.37	87.77	87.91	6.28			
62.45	62.59	4.47	75.18	75.32	5.38	87.91	88.05	6.29			
62.59	62.73	4.48	75.32	75.46	5.39	88.05	88.19	6.30			
62.73	62.87	4.49	75.46	75.60	5.40	88.19	88.33	6.31			
62.87	63.01	4.50	75.60	75.74	5.41	88.33	88.47	6.32			
63.01	63.15	4.51	75.74	75.88	5.42	88.47	88.61	6.33			
63.15	63.29	4.52	75.88	76.02	5.43	88.61	88.75	6.34			
63.29	63.43	4.53	76.02	76.16	5.44	88.75	88.89	6.35			
63.43	63.57	4.54	76.16	76.30	5.45	88.89	89.03	6.36			

Wages	Taxes
100	$7.15
200	14.30
300	21.45
400	28.60
500	35.75
600	42.90
700	50.05
800	57.20
900	64.35
1,000	71.50

PROBLEM

The first 2 lines of the Payroll Record have been completed properly; the next 18 lines remain to be completed. Instead of actually writing on the payroll record in this book, however, use another piece of paper. Make column headings similar to those on the Payroll Record, and fill in the correct information opposite each employee's number. Follow these steps:

1. Add the Total Wages and the Tips Reported to find the Total Wages Plus Tips.

2. Look at the proper income Tax Withholding Table, either the one for single employees, or the one for married employees, according to what is shown for that employee in the Payroll Record.

    Find the amount of tax to be withheld, REMEMBER—in the restaurant business, tips are added to wages to determine the proper withholding and social security deductions. Therefore, you use the figure from the Payroll Record under "Total Wages Plus Tips," and look in the Tax Withholding Table under "And the wages are—."

3. Look at the Social Security Employee Tax Tables to find the Social Security "Tax to be withheld" for the Total Wages Plus Tips.

    These tables show the tax on any amount up to $100. For wages over $100, look at the box in the lower right-hand corner of the tables; add the tax for the even hundred dollar amount to the figure from the tables above for the part that is less than $100.

    For example, for the first employee on the payroll record, who has total wages plus tips of $306.25, the figures are as follows:

    on $300.00 the tax to be withheld is $21.45
    on   6.25 the tax to be withheld is   .45
    on the total $306.25 the tax to be withheld is $21.90

4. Find the Total Deductions by adding the Withholding Tax figure and the Social Security Tax figures.

5. Find the Net Pay by subtracting the Total Deductions from the Total Wages.

LOCAL TAXES—HOW BUSINESS HELPS CARRY THE LOAD

Who pays for the policemen, firemen, libraries, parks, schools, street cleaners, and all the other public services in your city or town? The people who live there do. They pay taxes. People who do not live there, but work there or spend money there, also help out.

The governments of counties, cities, towns, and townships are called local governments. These governments throughout the United States collect many different kinds of taxes to pay their policemen, firemen, school teachers, and other public workers. Some of them collect income taxes, some collect sales taxes, some tax motor vehicles, but most of their money, however, comes from property taxes.

Property Taxes

Paying property taxes means paying a certain amount of money each year depending upon how much land and buildings and personal property you own. Property taxes are a favorite way for local governments to collect tax money because it is felt that the people who own the property receive many of the benefits of local government. Also, it is felt that the people who own property can probably afford to pay taxes better than those who do not own anything.

Everyone does not agree with this idea, of course, and that is why some local governments collect other types of taxes also, such as sales taxes. Thus, people who do not own property help pay the expenses of government.

Tax Rate. Each unit of local government, such as the city, the county, and the school district, plans a budget for the next year. Each unit sets a tax rate that will collect enough money from all the property owned in the unit to pay its expenses for the year. The total of these individual rates makes one combined rate that property owners pay.

Assessed Valuation. The county assessor determines the value of each person's property. For tax purposes, this value, called the assessed valuation, is usually about one-fourth of the actual value that the property could be sold for.

Property tax rates may be stated as percentages, or as so many dollars per thousand dollars of assessed valuation, or as so much per every hundred dollars of assessed valuation, or in some other similar way.

How Taxes Are Figured. Here are sample figures to show how a property tax rate works when the tax rate is $8.675 per hundred:

Actual value of house if sold = $80,000
Assessed valuation (one-fourth of market value) = $20,000
Tax rate = $8.675 per hundred dollars of assessed valuation

$20,000 ÷ 100 = 200

$8.675 × 200 = $1,735.00

The property taxes for the year would be $1,735.00.

Note that in using the term "$8.675 per hundred," we divide the assessed valuation by 100. If we use 8.675% of the assessed valuation as the tax rate, we change it to a decimal for multiplication by moving the point two places to the left, .08675, which is the same as dividing by 100. Therefore—$8.675 per hundred dollars of assessed valuation is the same as

8.675%   or   .08675 of assessed valuation

The $8.675 tax rate shown above would be the combined rate for all of the various taxing units within the county. The county tax collector would collect the money from the property owners and give each of the various taxing units, such as the city or the school district, its proper share.

Where Taxes Go

Most of the local taxes collected are used for schools. In one typical community it was found that there is an average of 1½ children per house. So, the more houses in a community, the more children there tend to be and the more schools are needed. Also, more children mean more policemen to protect them, more library books, more parks, playgrounds, etc.

Businesses need local government services, such as fire and police protection, good roads, etc.; but their needs are not as great or expensive as those for homes and families.

The problem is that the taxes collected from homes averaging 1½ children per home are usually not enough to pay for all the services needed in the community. Thus, it is very important for most communities to have some kind of business located there to help share the tax burden.

Here is an illustration which may help explain how business helps a community. Let us say the tax rate is 9% of the assessed valuation of all property.

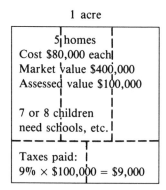
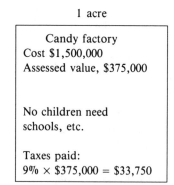

Notice the difference in income the local government receives from property taxes from the candy factory, $33,750, compared with that received from the five homes on the same size property, $9,000. However, the government spends less money providing services for the business than it does for the homes.

One very good business for a community to have is a hotel. A hotel may cost 5 million dollars or more to build, which means it would have an assessed valuation of a million dollars or more. It pays property taxes equal to 50 or more homes, but requires fewer government services than the homes.

In addition to property taxes, some local governments collect room taxes from hotels. That is, they collect a tax of as much as 5 percent of every dollar the hotels take in for renting rooms. This brings a large amount of money to the local government with very little additional expense.

PROBLEMS

I. Below are ten groups of tax districts. Find the combined tax rate for each group by adding them.

1. School, 2.81%; city, 1.9%; county, .9%

2. School, 2.935%; city, 2.15%; sanitary, 1.1%; county, .93%

3. School, 3.155%; city, 2.38%; lighting, .75%; county, 1.835%

4. School, 2.79%; city, 1.85%; water, 1.325%; county, .95%

5. School, 3.455%; city, 2.053%; sanitary, 1.3%; county, .9%

6. School, 3.875%; city, 2.14%; county, 2.67%

7. School, 3.459%; city, 1.875%; sanitary, 1.45%; county, 1.255%

8. School, 4.12%; city, 2.085%; lighting, 1.18%; county, .985%

9. School, 3.8725%; city, 1.54%; flood control, 1.13%; county, 1.275%

10. School, 4.274%; city, 1.786%; sanitary, 1.09%; county, 1.17%

11. School, 3.559%; county, 2.4%; mosquito abatement, 0.0112%

12. School, 3.7655%; city, 2.658%; county, 1.66%; hospital bonds, 0.0645%

13. School, 2.97%; city, 0.8472%; county, 1.231%; flood control, 0.3829%

14. School, 3.768%; city, 1.9762%; county, 1.905%; hospital bonds, 0.0012%

15. School, 4.5012%; city, 2.062%; county, 0.8796%; library bonds, 0.2281%; mosquito abatement, 0.0065%

II. Find the amount of taxes each of the following property owners pays by multiplying the assessed valuation by the combined tax rate given.

1. Assessed valuation, $7,950; total tax rate, 9.65%

2. Assessed valuation, $8,365; total tax rate, 8.955%

3. Assessed valuation, $10,050; total tax rate, 8.785%

4. Assessed valuation, $9,674; total tax rate, 9.853%

5. Assessed valuation, $14,532; total tax rate, 10.05%

6. Assessed valuation, $12,324; total tax rate, $7.76 per hundred

7. Assessed valuation, $38,952; total tax rate, $7.74 per hundred

8. Assessed valuation, $52,428; total tax rate, $9.625 per hundred

9. Assessed valuation, $132,624; total tax rate, $9.783 per hundred

10. Assessed valuation, $21,784; total tax rate, $12.1825 per hundred

11. Assessed valuation, $48,000; total tax rate, 7.9626%

12. Assessed valuation, $12,500; total tax rate, $9.755 per hundred

13. Assessed valuation, $10,500; total tax rate, 10.0765%

14. Assessed valuation, $13,750; total tax rate, $9.7066 per hundred

15. Assessed valuation, $12,225; total tax rate, 8.983%

III. Following are the assessed valuations of some pieces of business property. Approximately how many homes assessed at $7,500 each would be needed to equal the assessed valuation of each of these business properties? (Divide the assessed valuation of each by $7,500.) (Round off your answers to whole numbers.)

1. Restaurant, $21,684

2. Bowling alley, $34,852

3. Motel, $68,540

Approximately how many homes with an assessed valuation of $8,250 would be needed to pay the taxes paid by each of the following businesses?

4. Hotel, $258,462

5. Restaurant, $98,228

6. Clothing store, $24,085

Approximately how many homes with an assessed valuation of $8,750 would be needed to pay the taxes paid by each of the following businesses?

7. Auto dealership, $41,896

8. Hotel, $385,500

9. Grocery store, $62,114

10. Motel $128,600

QUESTIONS

1. Who pays for policemen, firemen, school teachers, and other public workers?

2. What is a local government?

3. What are property taxes?

4. What is assessed valuation?

5. What is the biggest expense for local governments?

6. How does having a hotel in the community help with the tax load?

TYPES OF BUSINESS OWNERSHIP

Who owns the restaurant you ate in the other day? Who owns the biggest hotel in town? Who owns the giant auto companies and steel mills? And who owns the acres of farms you see when you are out in the country?

Small Business vs. Large

Who owns any business depends largely on the size of the business. If it is a small business, it could be owned by one person. This is called single proprietorship. Proprietorship means ownership. If the business is too big for one person to own, it might be a partnership of two or three or more people. If it is a huge business, worth millions of dollars, it is probably a corporation.

There is another form of business ownership which is similar to a corporation, but which has a couple of important differences. It is called a co-operative. Finally, there is one other form of ownership which every country, including the United States, has and that is government ownership. The government owns such businesses as the postal system, national parks, some highways and bridges, military bases, etc.

SINGLE PROPRIETORSHIP

A single proprietorship means that there is just one owner of the business. If it makes money, the owner takes all the profits. If it loses money, the loss is all the owner's.

PARTNERSHIP

A partnership means ownership by two or more people. They make a legal agreement, preferably in writing, to operate a business as partners for a certain length of time. They agree on how much each partner is to invest in the business, what the duties of each is to be, and how profits and losses are to be shared. They don't have to be shared equally. The partners can agree on anything they wish.

CORPORATION

A corporation is a business that is owned by people called stockholders. There are usually a number of stockholders; but some individuals, such as doctors and other professional people, wish to take advantage of the laws applying to corporations, and they form a corporation with themselves as the only stockholders. People forming corporations must obtain a charter, that is, permission to incorporate from the state in which they locate their business.

Once the charter is granted, the corporation becomes a legal person itself, separate from the owners. In other words a corporation can do things people can do, such as enter into contracts and sue and be sued. If someone sues a corporation, he or she is not suing any person in the corporation but rather is suing the company itself.

How a Corporation Operates

The people who wish to organize a corporation agree to buy some stock in it. Buying stock means paying for shares of ownership. After they have been granted a charter, the owners sell shares of stock to other people whenever more money is needed to operate the business.

Each person who buys stock has one vote for each share of stock owned.* The stockholders call meetings at least once a year. Using their one vote for each share they own, they cast votes to elect a board of directors.

The board of directors appoints officers of the corporation, usually a president, one or more vice-presidents, a secretary, and a treasurer.

The board of directors also meets frequently to make major decisions. In between board meetings, the officers run the company. They hire the other employees and make daily decisions.

The board of directors decides how to use the profits, if there are any. They may decide to keep all the profits in the business to help it grow. Or they may decide to keep part of the profits in the business and pay some of the profits to the stockholders.

Corporation Dividends. Paying profits to the stockholders is known as paying dividends. The dividends are distributed equally among the shares of stock owned. The dividend, for example, might be 50 cents per share. A person owning 50 shares would get $25, and a person owning 1,000 shares would get $500.

Corporation Losses. If there are losses, the corporation cannot ask the stockholders to pay in more money to cover the losses. The stockholders can only lose what they originally paid for the stock. It could become worthless if the company were to go broke. Additional shares of stock could be sold, however, to bring more money into the corporation.

CO-OPERATIVE

A co-operative is another form of business ownership which must get a charter from the state. In a co-operative a group of people get together to buy in larger quantities or to sell in larger quantities than they could as individuals. Farmers frequently form co-operatives for buying feed and other farm supplies. In this way they can buy more cheaply. They also combine products they want to sell. In selling one large amount of a product as a group, they have more power in dealing with the buyer than if they each sold individually.

High schools and colleges frequently have co-operative book stores. The students who buy there are members. Any profits from the books are returned to the students by making the books cheaper or by giving refunds to the students at the end of the year.

---

* The stock referred to here is technically known as "common stock," as opposed to another type of stock known as "preferred stock," which has preference in collecting dividends but has no voting privileges. Most companies sell only common stock.

ADVANTAGES AND DISADVANTAGES OF DIFFERENT FORMS OF BUSINESS ORGANIZATION

Advantages | *Disadvantages*

Single Proprietorship

Owner has complete control. He is his own boss.

He gets the benefits of his hard work and good ideas and good reputation.

All profits are his.

The responsibility and worries are all his. There is no one to really share them with.

Limited capital investment. The only money that is invested in the business is the money he has plus what he can borrow.

All the losses are his.

Partnership

More talents are available to the business. Two heads are sometimes better than one. One might be an excellent cook and the other an excellent accountant. Each contributes his skills for a better business.

Two or more people can invest more money than one person.

Two or more people can borrow more money than one person.

Two or more people can share the duties of being boss, especially if the business runs for more than eight hours a day.

If there are losses, there is someone else to share them.

Partners do not always agree with each other on important decisions.

Each partner is legally responsible for the actions of the other partners in the business. If the partnership business loses money and one partner can not pay his share of the losses, the other partners must pay his share from their personal bank accounts.

Limited life of business: the partnership only lasts as long as the agreement states. If one partner dies, the partnership stops and some arrangement has to be made to give the dead partner's relatives their share of the business.

If there are profits, there is someone else to share them with.

Corporation

Transfer of ownership. If an owner dies, the corporation is not affected. Someone inherits the stock but the corporation continues as before except that it has some new owners.

Ease of raising money to invest in the business: It is easier to sell 100,000 shares of stock at $10 each to get a million dollars than to find one person or a few partners with a million dollars to invest.

Expert leadership. A corporation can afford to hire experts for various jobs in the business.

Limited liabilities: The people who buy stock can only lose the money they pay for the stock. A single proprietor or partners can lose their personal bank accounts to pay for business losses.

Difficulty of starting: It is more complicated to start a business as a corporation than as a single proprietorship or a partnership.

Stockholders can be out-voted on selecting the board of directors and on other major decisions by people who own more shares of stock.

Double taxation: Since the corporation is the same as a person legally, it pays income taxes on its profits. When the stockholders receive the profits in the form of dividends, they are taxed again as they are individual taxpayers. In that way the profits of the business have been taxed twice.

Co-operative

Money can be saved by buying in large quantities, and better deals can be made in selling in larger quantities.

Requires a charter from the state.

Requires people to run it.

QUESTIONS

1. Small businesses usually operate under what type of ownership?

2. A million-dollar business usually operates under what type of ownership?

3. Define briefly each of the following terms:
   a. Single proprietorship
   b. Partnership
   c. Corporation
   d. Co-operative

4. How many votes does each stockholder get in running a corporation?

5. What are dividends?

6. In what business or place are co-operatives often found?

7. What is one serious disadvantage of a partnership?

8. Why do most large businesses use the corporation type of organization?

9. What is a big advantage of investing money in a corporation?

10. What is a big disadvantage of investing money in corporations?

PLANS FOR AN INN— A SCALE DRAWING

When a group of business people decide that they might like to build a hotel or a motel, also called an inn, they have plans drawn up by an architect. These plans are known as scale drawings. A scale drawing, or plan, is one in which the drawing is quite a bit smaller than the actual building is going to be, but it is in perfect proportion. For example, 1/4 inch on the plan could be made equal to 1 foot of the actual building—that would be a scale of 1/4" = 1'.

On the scale drawing for the Feast Inn pictured below, each 1 inch on the plan is equal to 40 feet of the actual inn. That is a scale of 1" = 40'.

Scale: 1" = 40'

PROBLEMS

Below is a list of eighteen areas of the inn. On the first two lines you will see the measurements of the scale drawing and the actual measurements and square feet in each area when the inn is built. Complete the remaining sixteen lines as follows:

a. Convert the measurements of the drawing to the actual length and width of each area. To do this, multiply the number of inches by forty and record the answer in feet. For example, the length of the manager's office on the plan measures three-fourths of an inch.

$$\frac{3}{4} \times \frac{40}{1} = 30 \text{ feet actual length}$$

b. Find the square feet of each area by multiplying the actual length by the actual width.

	Length in Inches on Drawing	Width in Inches on Drawing	a — Actual Length in Feet	Actual Width in Feet	b. Area in Square Feet
1. Manager's office	3/4"	1/4"	30'	10'	300 sq. ft.
2. Front office	3/4"	1/2"	30'	20'	600 sq. ft.
3. Entrance lobby	1 3/4"	1/4"			
4. An outside room	3/8"	3/8"			
5. Each corridor	3 3/4"	1/4"			
6. Each exit	3/8"	1/4"			
7. A poolside room	11/16"	3/8"			
8. A housekeeper's room	11/16"	3/8"			
9. Engineer's Department	15/16"	1/2"			
10. Storeroom	15/16"	1/4"			
11. Cocktail lounge	3/4"	1/2"			
12. Dining Room	1 1/8"	3/4"			
13. Kitchen	3/4"	7/16"			
14. Conference Room A	3/4"	5/8"			
15. Conference Room B	3/4"	11/16"			
16. Rear entrance	1 5/16"	1/4"			
17. Swimming pool	1 9/16"	3/4"			
18. Total property, Feast Inn	5 1/8"	4 3/4"			

WHAT EMPLOYERS LOOK FOR WHEN HIRING

Your appearance makes the first impression, and cleanliness and neatness are the keys to a good appearance.

CHECK LIST FOR A GOOD APPEARANCE FOR THE JOB INTERVIEW

Shower or bath in the morning.
Teeth brushed. Clean breath.
Hair combed neatly. (Length and style depend upon the kind of establishment you are applying to. Notice how their present employees wear their hair.)
An alert, healthy face—the result of good health and a good night's rest the night before the interview.

Young Men

Clean hands and fingernails.
Jacket and tie. (Not always necessary, but it sure improves your chances of winning the job over competitors.)
Shoes polished.

Young Women

Clean hands and manicured nails. Conservative color in nail polish.
Business-like dress, skirt, or pants, blouse, or sweater.
Hose—natural colors, no snags or runs.
Dressy shoes, polished.
Not too much make-up or jewelry.
Slip or other undergarments not showing. (If you can, check your appearance in a full-length mirror.)
No dandruff or stray hairs on your shoulders.

OTHER TIPS FOR JOB APPLICANTS

Apply for the job alone. Don't bring along a friend. Introduce yourself. Don't wait for the employer to ask your name. Say your name clearly. Don't chew gum or smoke during the interview unless invited to. If you have an appointment, be prompt. (Plan to get there 10 minutes early.) Be prepared to answer questions normally asked during an employment interview, such as:

1. Why do you want to work here?

2. What salary do you expect? (Find out the union wage scale before you apply.)

3. What experience have you had? (Be able to describe your duties on previous jobs.)

4. Why did you leave your last job—were you fired?

5. Tell me about yourself. (This may be simply to hear you speak. Watch your grammar.)

Fill out written application and forms neatly and completely. Know names and addresses of former employers and personal references. Bring your social security card with you.

And don't be afraid—the employer is just as anxious to find a good worker as you are to find a good job. He asks you questions and looks you over because he is trying to find the person who will fit the job and do well at it.

Attitude is really important in the hospitality business. *Smile— and mean it!*

QUESTIONS

1. What are the two keys to a good appearance?

2. How can you determine what hair styles the employer likes? (He might be too polite to tell you he does not approve of yours, and he may give some other reason for not hiring you.)

3. What types of hose do businesspeople prefer their female employees to wear?

4. How do employers feel about make-up and jewelry on girls?

5. Is it wise to have a friend with you when you apply for a job?

6. What are three questions, at least, that employers are sure to ask you when you apply for work?

7. Why do employers ask job applicants so many questions?

ACTIVITY PROBLEMS

Following are "activity problems," which may not seem like problems. After you have read one of these problems, you may even feel you have to restate the problem in your own terms before you attempt a solution.

You may be evaluated in many ways. The following questions may be helpful in evaluation.

1. Is the restatement of the problem logical?
2. Has the problem been well thought out?
3. Is the method you are trying a step toward solving the problem? (This does not mean that your method of attack has to be the same as anyone else's.)
4. Are you able to defend your method of attack in case of criticism?
5. Is your solution based on your own good thinking and/or sound research?

Activity Problem I. You make out the payroll checks for a restaurant that has 37 employees. All employees except the manager punch time cards. One Friday the restaurant is broken into and, among other things, the time cards for that week are missing.

Activity Problem II. Your restaurant burns to the ground one night.

Activity Problem III. Design a motel-restaurant-bar complex on a suitable site with the desired goal that it will return $35,000 a year income to the owner.

Activity Problem IV. Measure the area and volume of your classroom, or some other room or building.

Activity Problem V. Find out how you determine the kind and size of an individual space heater needed for an average-size hotel room in your area.

Activity Problem VI. Study the layout of a local restaurant kitchen. Redesign the kitchen to make it more efficient. Invite the owner or manager to the classroom to hear your presentation.

Activity Problem VII. Work out menus for the seven days of the week, and justify them in terms that

1. they satisfy and provide for good nutrition standards,
2. they will be salable to the public, and
3. they provide meals which will return at least 30% profit to the establishment.

Activity Problem VIII. If the moon space ship "Columbia" was on automatic control from Lift-off -6 hours to lift-off $+27$ hours, how long was the space ship on automatic control?

Draw a graph, using a number line, to illustrate the problem. Then name and illustrate a similar problem that could be encountered in the foodservice industry.

Activity Problem IX. Draw up a complete budget for a family of four—a father, 37 years old, a mother, 36, a daughter, 15, and a son, 9. Use a circle graph to illustrate percentages and categories. Present and defend your budget before the class.

SUPPLEMENTARY TABLES AND CHARTS

Look through this section carefully, one page at a time. Gold can be hidden on some of the following pages. The data collected in these tables and charts can supply a wealth of information. However, you must know how to find and use the tables and charts in order to unlock valuable information. Study carefully all directions and practice using some of the charts. For example, by courtesy of the Koch Company, some valuable charts are included that make for quick and easy figuring of cost per portion. The first chart has a section that tells how to use the charts. Read the directions, learn how to use the charts, and many hours of valuable time may be saved.

TABLES OF MEASURE

Dry Measure

2 pints = 1 quart
8 quarts = 1 peck
4 pecks = 1 bushel

Long Measure

12 inches = 1 foot
3 feet = 1 yard
5½ yards = 1 rod
320 rods = 1 mile
5280 feet = 1 mile
1760 yards = 1 mile

Cubic Measure

1728 cubic inches = 1 cubic foot
27 cubic feet = 1 cubic yard
128 cubic feet = 1 cord
1¼ cubic feet = 1 bushel (approx.)
.8 bushels = 1 cubic foot (approx.)
7½ gallons = 1 cubic foot (approx.)
2150.42 cubic inches = 1 bushel
231 cubic inches = 1 gallon

Time Measure

60 seconds = 1 minute
60 minutes = 1 hour
24 hours = 1 day
7 days = 1 week
30 days = 1 month
365 days = 1 year
52 weeks = 1 year
12 months = 1 year

Liquid Measure

4 gills = 1 pint
2 pints = 1 quart
4 quarts = 1 gallon

Square Measure

144 square inches = 1 square foot
9 square feet = 1 square yard
30¼ square yards = 1 square rod
160 square rods = 1 acre
640 acres = 1 square mile

Avoirdupois Weight

16 ounces = 1 pound
100 pounds = 1 cwt. (hundred weight)
2000 pounds = 1 net (short) ton
2240 pounds = 1 gross (long) ton

Miscellaneous

A.P. means As Purchased
E.P. means Edible Portion

Note: The legal definition of a month is accepted as 30 days, even though the actual month may have more or fewer days.

A common year contains 365 days. A leap year contains 366 days and occurs at intervals of 4 years; a year evenly divisible by 4 (such as 1492, 1776, 1812, 1960) is a leap year. However, leap year occurs at the end of a century only when the year is evenly divisible by 400, e.g. 800, 1200, 1600, etc.

EGG SPECIFICATIONS

Since eggs are among those products whose characteristics of quality undergo very rapid changes, it should be kept in mind when buying eggs that the grades indicate the quality at the time of grading. Each grade stamp will indicate the date of grading. The four official grades of quality are: AA, A, B, and C. These grades are determined by the size, exterior and interior appearance of the eggs. All grades come in all sizes, and the size is an important factor in price.

The size classes of eggs are divided as follows:

Size Class	Minimum Net Weight (in pounds) 30-Dozen Crate
Jumbo	52
Extra Large	48½
Large	45
Medium	40
Small	34

The grading of the egg's quality depends on the clearness of the white, blemishes of yolk, size of the air cell within the egg, and the condition of the shell.

It may be seen that there is a 3 ounce difference between classes. It is easy to remember by establishing large as 24 ounces and adding or subtracting 3 ounces in steps to determine proper weights for other classes.

EGG GRADES

The quality of eggs depends on flavor and appearance. Eggs are graded according to appearance and size. There are four U. S. grades of quality: AA, A, B, and C. AA are the finest quality. Grade A eggs are suitable for all table use. Grade B eggs are primarily for baking. Grade C eggs are strictly for baking.

U.S. WEIGHT CLASSES, CONSUMER GRADES FOR EGGS IN THE SHELL

Size or Weight Class	Minimum Net Weight per Dozen
Jumbo	30 oz.
Extra Large	27 oz.
Large	24 oz.
Medium	21 oz.
Small	18 oz.
Peewee	15 oz.

Koch Food Service Bulletins

Prepared by the Research and Educational Department of Koch Refrigerators, Inc.

The use of standardized recipes is an important factor in portion control. A recipe, however, can be depended upon to give the stated number of portions only if the servings are of a uniform size. The most dependable method to use when measuring portions is to serve the food with ladles, scoops, and spoons of standard size.

LADLES

Ladles may be used in serving soups, creamed dishes, stews, sauces, gravies, and other similar products.
The following sizes of laddles are most frequently used for serving:

1/4 cup (2 ounces)
1/2 cup (4 ounces)
3/4 cup (6 ounces)
1 cup (8 ounces)

SCOOPS

The number of the scoop indicates the number of scoopfuls it takes to make one quart. The table below shows the level measures of each scoop in cups or tablespoons.

Scoop Number	Level Measure
6	2/3 cup
8	1/2 cup
10	2/5 cup
12	1/3 cup
16	1/4 cup
20	3 1/5 tablespoons
24	2 2/3 tablespoons
30	2 1/5 tablespoons
40	1 3/5 tablespoons

Scoops may be used for portioning such items as drop cookies, muffins, meat patties, and some vegetables, salads, and sandwich fillings.

SERVING SPOONS

A serving spoon (solid or perforated) may be used instead of a scoop. Since these spoons are not identified by number, it is necessary to measure or weigh food in the spoons used to obtain the approximate serving size desired.

FRACTIONAL EQUIVALENTS FOR USE IN CONVERTING RECIPES

The following chart is designed to help you change fractional parts of pounds, gallons, cups, etc., to accurate weights or measures. For example, reading from left to right, the table shows that 7/8 of one pound is 14 ounces, 1/3 of a gallon is 1 quart plus 1⅓ cups, 1/16 of a cup is 1 tablespoon; etc.

	1 Tablespoon	1 Cup	1 Pint	1 Quart	1 Gallon	1 Pound
1	3 tsp.	16 Tbsp.	2 cups	2 pints	4 quarts	16 ounces
7/8	2½ tsp.	1 cup less 2 Tbsp.	1¾ cups	3½ cups	3 quarts plus 1 pint	14 ounces
3/4	2¼ tsp.	12 Tbsp.	1½ cups	3 cups	3 quarts	12 ounces
2/3	2 tsp.	10 Tbsp. plus 2 tsp.	1⅓ cups	2⅔ cups	2 quarts plus 2⅔ cups	10⅔ ounces
5/8	2 tsp. (scant)	10 Tbsp.	1¼ cups	2½ cups	2 quarts plus 1 pint	10 ounces
1/2	1½ tsp.	8 Tbsp.	1 cup	2 cups	2 quarts	8 ounces
3/8	1⅛ tsp.	6 Tbsp.	3/4 cup	1½ cups	1 quart plus 1 pint	6 ounces
1/3	1 tsp.	5 Tbsp. plus 1 tsp.	2/3 cup	1⅓ cup	1 quart plus 1⅓ cups	5⅓ ounces
1/4	3/4 tsp.	4 Tbsp.	1/2 cup	1 cup	1 quart	4 ounces
1/8	1/2 tsp. (scant)	2 Tbsp.	1/4 cup	1/2 cup	1 pint	2 ounces
1/16	1/4 tsp. (scant)	1 Tbsp.	2 Tbsp.	4 Tbsp.	1 cup	1 ounce

CONTAINER PORTION AND CONVERSION CHART

Fluid Ounces	Cups	No. 8 Scoops	No. 12 Scoops	Containers	No. 16 Scoops	No. 24 Scoops	No. 30 Scoops	No. 40 Scoops
130	16	32	48	ONE GALLON	64	96	120	160
					62	93		
120	15	30	45		60	90		150
					58	87		
110	14	28	42		56	84	105	140
					54	81		
	13	26	39		52	78		130
100				ONE #10 CAN	50	75		
	12	24	36	THREE QUARTS	48	72	90	120
				TWO #5 CANS	46	69		
90	11	22	33		44	66		110
					42	63		
80	10	20	30	THREE #2½ CANS	40	60	75	100
					38	57		
70	9	18	27		36	54		90
					34	51		
	8	16	24	ONE HALF GALLON	32	48	60	80
60					30	45		
	7	14	21		28	42		70
				TWO #2½ CANS	26	39		
50	6	12	18	ONE #5 CAN	24	36	45	60
					22	33		
40	5	10	15		20	30		50
					18	27		
	4	8	12	ONE QUART	16	24	30	40
30				ONE #2½ CAN	14	21		
	3	6	9		12	18		30
20					10	15		
	2	4	6	ONE #303 CAN	8	12	15	20
8	1	2	3	ONE CUP	4	6	7½	10

CONVERSION TABLES

	LOGS
1 ATMOSPHERE =	
76 cms. of mercury	1.88081
29.92 ins. of mercury	1.47596
33.90 ft. of water	1.53020
10333 kgs. per sq. m.	4.01424
14.70 lbs. per sq. in.	1.16732
1013200 dynes per sq. in.	6.00570
1 ACRE =	
square with sides 208.71 ft.	
43560 sq. ft.	4.63909
4046.9 sq. m.	3.60712
160 sq. rods	2.20412
4840 sq. yds.	3.68485
.40469 hectares	9.60712*
1 ACRE FOOT =	
43560 cu. ft.	4.63909
325850 gals.	5.52162
1 ACRE FT. PER DAY =	
.5 sec. ft.	9.69897*
1 BARREL (LIQUID) =	
31.5 gals.	1.49831
1 B. T. U. =	
252.2 gm. cal.	2.40175
777.64 ft. lbs.	2.89078
107.6 kg. m.	2.03181
1055 joules	3.02325
.293 watt hours	9.46538*
1 B. T. U. PER MIN. =	
.02356 H. P.	8.37218*
17.57 watts	1.24477
1 B. T. U. PER SEC. =	
1.4138 H. P.	0.14039
1052.6 watts	3.02227
1 BUSHEL =	
1.244 cu. ft.	0.09482
.0354 cu. m.	8.54900*
2150.42 cu. ins.	3.33252
32 dry quarts	1.50515
4 pecks	0.60206
35.2383 litres	1.54702
1 CALORIE (GM)	
.003965 B. T. U.'s	7.59824*
.00000 162 kw. hrs.	4.20952*
4.183 joules	0.62149
3.087 ft. lbs.	0.48954
.4267 kg. M.	9.63012*

1 CENTIMETER =	
.3937 ins.	9.59517*
.01 meters	8.00000*
1 SQ. CENTIMETER =	
.155 sq. ins.	9.19033*
100 sq. mm.	2.00000
1 CU. CENTIMETER =	
.06102 cu. ins.	8.78547*
1000 cu. mm.	3.00000
1 CENTIMETER PER SEC. =	
.036 km. per hr.	8.55630*
.02237 mi. per hr.	8.34967*
1 CHEVAL VAPEUR =	
75 kg. m. per sec.	1.87506
.7355 kws.	9.86659*
.9863 H. P.	9.99402*
1 CIRCULAR INCH =	
.7854 sq. ins.	9.89509*
1 DEGREE =	
.01745 radians	8.24180*
1 DEGREE PER SEC. =	
.1667 R. P. M.	9.22185*
1 DAY =	
24 hours	1.38022
1440 mins.	3.15836
86400 secs.	4.93651
1 DYNE =	
.0010204 gms.	7.00089*
.000002247 lbs.	4.35160*
.0000 7233 poundals	5.85932*
1 FATHOM =	
6 ft.	0.77815
1 FOOT =	
.01515 chains	8.18041*
30.48 cms.	1.48401
.001515 furlongs	7.18041*
12 ins.	1.07918
.304801 meters	9.48401*
.06061 rods	8.78254*
.0001894 miles	6.27738*
.3333 yds.	9.52288*
1 SQ. FOOT =	
.00002296 acres	5.36097*
144 sq. ins.	2.15836
.0929 sq. meters	8.96803*
.003673 sq. rods	7.56503*
.1111 sq. yds.	9.04576*

1 CU. FOOT =	LOGS
28316.8 c. c.	4.45205
1728. cu. ins.	3.23754
.037037 cu. yds.	8.56846*
.028317 cu. meters	8.45211*
7.481 gals.	0.87396
28.32 litres	1.45209
1 FOOT POUND =	
.001285 B. T. U.'s	7.10890*
13563000 ergs	7.13231
3241 gm. cal.	3.51068
1.356 joules	0.13226
.13826 Kg. M.	9.14068*
1 FOOT POUND PER MIN =	
.0000303 H.P.	5.48144*
.0000226 Kws.	5.35411*
1 FOOT POUND PER SEC. =	
.001818 H. P.	7.25959*
.001356 Kws.	7.13226*
1 FOOT PER MIN. =	
5080 cms. per sec.	3.70586*
.01829 km. per hr.	8.26221*
.01136 mi. per hr.	8.05538*
1 FOOT PER SEC. =	
1.0973 km. per hr.	0.04032
68182 mi. per hr.	9.83367
1 FOOT OF WATER =	
.0295 atmospheres	8.46982*
.8826 in. of mercury	9.94576*
1 FURLONG =	
660 ft.	2.81954
40 rods	1.60206
10 chains	1.00000
220 yards	2.34242
1 GALLON =	
1337 cu. ft.	9.12613*
4 quarts	0.60206
3.78526 litres	0.57811
231 cu. ins.	2.36361
8.3356 lbs. of water	0.92094
1 GRAM =	
980 dynes	2.99123
.0022046 lbs.	7.34333*
1 (approx) c.c. of water.	
1 GRAM PER C. C.	
62.43 lbs. per cu. ft.	1.79539

G. (GRAVITY) =	
980 cms. per sec.^{2}	2.99123
32.2 ft. per sec.^{2}	1.50786
1 HECTARE =	
Square with 100M Sides.	
2.471 acres	0.35162
1 HORSE-POWER =	
42.44 B. T. U.'s per min.	1.62777
.707 B. T. U.'s per sec.	9.84942*
1070 gm. cal. per min.	3.02938
33000 ft. lbs. per min.	4.51851
550 ft. lbs. per sec.	2.74036
.746 kw.	9.87274*
746 watts	2.87274
1 H.P. BOILER) =	
33520 B. T. U.'s per hr.	4.52530
9.805 kw.	0.99140
1 HOUR =	
60 min.	1.77815
3600 sec.	3.55630
1 INCH =	
.8333 ft.	9.92082*
.02778 yds.	8.44373*
.0505 rods	8.70329*
.00015785 mi.	6.19819*
2.54001 cms.	0.40483
1 SQUARE INCH =	
.006944 sq. ft.	7.84161*
6.452 sq. cms.	0.80969
1 CUBIC INCH =	
.0005787 cu. ft.	6.76245*
16.387 c.c.	1.21449
1 IN. MERCURY =	
.03342 atmospheres	8.52401*
.4912 pds. per sq. in.	9.69126*
1.133 ft. water	0.0543
13.6 in. water	1.13354
0.491 lbs. per sq. in.	9.69081*
345.3 kg. per sq. meter	2.53820
1 INCH OF WATER =	
.002458 atmospheres	7.39058*
.07355 in mercury	8.86658*
5.204 lbs. per sq. ft.	0.71634
.03613 lbs. per sq. in.	8.55787*

denotes − 10 . . . thus 7.10890 is logarithm .10890 with − 3 characteristic.
Studyplan Brochure, Bank of America, San Francisco.

	LOGS
1 JOULE =	
.0009486 B. T. U.'s	6.97708*
.2389 calories	9.37822*
10,000,000 ergs	7.00000
.7376 ft. lbs.	9.86782*
.1020 kg. ms.	9.00860*
1 KILOGRAM =	
980,000 dynes	5.99123
2.20462 lbs.	0.34333
2.6792 Troy lbs.	0.42800
1 KILOMETER =	
3280.83 ft.	3.51598
.6214 mi.	9.79337*
19.8838 rods	1.29850
1093.6 yds.	3.03862
1 SQ. KILOMETER =	
247.11 acres	2.39289
.386109 sq. miles	9.58670*
1,000,000 sq. meter	6.00000
1,196,000 sq. yds.	7.07773
1 KILOMETER PER HR. =	
.91134 ft. per sec.	9.99621*
27.78 cms. per sec.	1.44373
1 KILOWATT =	
56.92 B. T. U.'s per m.	1.75527
.949 B. T. U.'s per sec.	9.97727*
239 gm. cals. per sec.	2.37840
44240 ft. lbs. per min.	4.64582
737.3 ft. lbs. per sec.	2.86764
1.341 H. P.	0.12743
1000 watts	3.00000
1 KNOT =	
(Min. Longtidue at Equator)	
1.853 km. per hr.	0.26788
1.1516 mi. per hr.	0.06130
6080.26 ft. per hr.	3.78392
1 LINK (ENGINEER'S) =	
12 ins.	1.07981
1 LINK (SURVEYOR'S) =	
7.92 ins.	0.89873
.08 chains	8.00000*
1 LIGHT YEAR =	
5900 billion miles	12.77085
9500 billion kms.	12.97772

	LOGS
1 LITRE =	
.03531 cu. ft.	9.54790*
61.02 cu. in.	2.78675
33.8147 fl. ounces	1.52910
1000.027 c.c.	3.00001
.9081 dry quarts	9.95813*
.2642 gals.	9.42193*
1.057 qts.	0.02403
$LOG_{10}N = 2.303\ LOG_e$	0.35229
$LOG_eN = 2.343\ LOG_{10}$	9.63779
e = 2.7182818	0.43429
1 METER =	
100 cms.	2.00000
3.28083 ft.	0.51598
39.37 ins.	1.59517
.1988 rods	9.29842*
1.09361 yds.	0.03862
1 SQ. METER =	
10,000 sq. cms.	4.00000
10.764 sq. ft.	1.03197
1.196 sq. yds.	0.07773
.0002471 acres	6.39287*
1 CU. METER =	
1,000,000 c.c.	6.00000
35.314 cu. ft.	1.54795
1.308 cu. yds.	1.11661
264.2 gals.	2.42193
1 MILLIMETER =	
.03937 ins.	8.59517*
.1 cm.	9.00000*
1 SQ. MILLIMETER =	
.0015499 sq. ins.	7.19027*
1 CU. MILLIMETER =	
.000061 cu. ins.	5.78533*
1 MILE =	
5280 ft.	3.772263
8 furlongs	0.90309
1.60935 km.	0.20665
.8684 knot hrs.	9.93872*
320 rods	2.50515
1760 yds.	3.24551
1 SQ. MILE =	
640 acres	2.80618
27878400 sq. feet	7.44527
258999.8 sq. meters	5.41330
2.589998 sq. km.	0.41330
102400 sq. rods	5.01030
3097600 sq. yds.	6.49102

	LOGS
1 MILE PER HR. =	
1.46667 ft. per sec.	0.16633
44.70 cms. per sec.	1.65031
88 ft. per min.	1.94448
1 MINUTE (ANGULAR) =	
.0002909 radians	6.46374*
1 OUNCE' (AVD.) =	
16 drams	1.20412
437.5 grains	2.64098
28.35 grams	1.45255
.0625 lbs.	8.79588*
1 POUND (AVD.) =	
444,518.2 dynes	5.64787
453.59 gms.	3.65667
.45359 kgs.	9.65667*
27.692 cu. ins. water	1.44235
7000 grains	3.84510
256 avd. drams	2.40824
1453 troy oz.	3.16227
.1198 gal. water	9.07882*
1 POUND (TROY) =	
5760 grains	3.76042
.8229 lbs. (adv.)	9.91535*
12 oz. (Troy)	1.07918
240 pennyweights	2.38021
373.24177 grams	2.57199
1 POUNDAL =	
13826 dynes	4.14069
14.10 grams	1.14922
.03108 lbs.	8.49248*
1 POUND PER CU. FT. =	
.01602 grams per c.c.	8.20466*
1 QUART =	
.25 gals.	9.39794*
.9463 litres	9.97603*
946.3 c. c.	2.97603
57.75 cu. ins.	1.76155
2.0839 lbs. of water	0.31877
1 RADIAN =	
57.2958 degrees	1.75812
3438 minutes	3.53631
1 ROD =	
.25 chains	9.39794*
16.5 ft.	1.21748
.003125 miles	7.49485*

	LOGS
1 SQ. ROD =	
.00625 acres	7.79588*
272.25 sq. ft.	2.43497
25.29 sq. meters	1.40295
30.25 sq. yds.	1.48072
1 SECOND (ANGULAR) =	
.000004848 radians	4.68556*
.016667 minutes	8.22212*
.00027778 degrees	6.44370*
1 SECOND FT. =	
646300 gals. per day	5.81043
2 acre ft. per day	0.30103
1 SLUG =	
32.2 lbs.	1.50786
14.6 kilograms	1.16435
1 TON (SHORT) =	
2000 lbs.	3.30103
907.185 kgs.	2.95769
1 WATT =	
44.24 ft. lbs. per min.	1.64582
1 joule per sec.	
.001 kw.	7.00000*
WATER—1 CU. FT. =	
62.425 lbs.	1.79536
7.4805 gals.	0.87393
28.317 kg.	1.45204
WATER—1 CU. METER =	
264.17 gals.	2.42187
1000 kg.	3.00000
2204 lbs.	3.34322
1 YARD =	
3 ft.	0.47712
36 ins.	1.55630
.9144 meters	9.96114*
1 SQ. YARD =	
.0002066 acres	6.31515*
9 sq. ft.	0.95424
1296 sq. ins.	3.11261
.836 sq. meters	9.92221*
.03306 sq. rods	8.51930*
1 CU. YARD =	
46656 cu. ins.	4.66891
27 cu. ft.	1.43136
.7645 cu. meters	9.88338*

denotes −10 ... thus 7.10890 is logarithm .10890 with −3 characteristic.
Studyplan Brochure, Bank of America, San Francisco.

DECIMAL EQUIVALENTS OF COMMON FRACTIONS

1/4		1/6		1/8		1/12		1/16		1/32	
1	.25	1	.1667	1	.125	1	.0833	1	.0625	1	.03125
2	.5	2	.3333	2	.25	2	.1667	2	.125	2	.0625
3	.75	3	.5	3	.375	3	.25	3	.1875	3	.09375
		4	.6667	4	.5	4	.3333	4	.25	4	.125
		5	.8333	5	.625	5	.4167	5	.3125	5	.15625
				6	.75	6	.5	6	.375	6	.1875
				7	.875	7	.5833	7	.4375	7	.21875
						8	.6667	8	.5	8	.25
						9	.75	9	.5625	9	.28125
						10	.8333	10	.625	10	.3125
						11	.9167	11	.6875	11	.34375
								12	.75	12	.375
								13	.8125	13	.40625
								14	.875	14	.4375
								15	.9375	15	.46875
										16	.5
										17	.53125
										18	.5625
										19	.59375
										20	.625
										21	.65625
										22	.6875
										23	.71875
										24	.75
										25	.78125
										26	.8125
										27	.84375
										28	.875
										29	.90625
										30	.9375
										31	.96875

Friden Division, The Singer Co.

OUNCES AND EIGHTHS OF OUNCES EXPRESSED AS DECIMAL PARTS OF POUND

Ounces	1/8ths	Decimal Part of Pound		Ounces	1/8ths	Decimal Part of Pound		Ounces	1/8ths	Decimal Part of Pound		Ounces	1/8ths	Decimal Part of Pound	
0	0	.0000	000	4	0	.2500	000	8	0	.5000	000	12	0	.7500	000
	1/8	.0078	125		1/8	.2578	125		1/8	.5078	125		1/8	.7578	125
	1/4	.0156	250		1/4	.2656	250		1/4	.5152	250		1/4	.7656	250
	3/8	.0234	375		3/8	.2734	375		3/8	.5234	375		3/8	.7734	375
	1/2	.0312	500		1/2	.2812	500		1/2	.5312	500		1/2	.7812	500
	5/8	.0390	625		5/8	.2890	625		5/8	.5390	625		5/8	.7890	625
	3/4	.0468	750		3/4	.2968	750		3/4	.5468	750		3/4	.7968	750
	7/8	.0546	875		7/8	.3046	875		7/8	.5546	875		7/8	.8046	875
1	0	.0625	000	5	0	.3125	000	9	0	.5625	000	13	0	.8125	000
	1/8	.0703	125		1/8	.3203	125		1/8	.5703	125		1/8	.8203	125
	1/4	.0781	250		1/4	.3281	250		1/4	.5781	250		1/4	.8281	250
	3/8	.0859	375		3/8	.3359	375		3/8	.5859	375		3/8	.8359	375
	1/2	.0937	500		1/2	.3437	500		1/2	.5937	500		1/2	.8437	500
	5/8	.1015	625		5/8	.3515	625		5/8	.6015	625		5/8	.8515	625
	3/4	.1093	750		3/4	.3593	750		3/4	.6093	750		3/4	.8593	750
	7/8	.1171	875		7/8	.3671	875		7/8	.6171	875		7/8	.8671	875
2	0	.1250	000	6	0	.3750	000	10	0	.6250	000	14	0	.8750	000
	1/8	.1328	125		1/8	.3828	125		1/8	.6328	125		1/8	.8828	125
	1/4	.1406	250		1/4	.3906	250		1/4	.6406	250		1/4	.8906	250
	3/8	.1484	375		3/8	.3984	375		3/8	.6484	375		3/8	.8984	375
	1/2	.1562	500		1/2	.4062	500		1/2	.6562	500		1/2	.9062	500
	5/8	.1640	625		5/8	.4140	625		5/8	.6640	625		5/8	.9140	625
	3/4	.1718	750		3/4	.4218	750		3/4	.6718	750		3/4	.9218	750
	7/8	.1796	875		7/8	.4296	875		7/8	.6796	875		7/8	.9296	875
3	0	.1875	000	7	0	.4375	000	11	0	.6875	000	15	0	.9375	000
	1/8	.1953	125		1/8	.4453	125		1/8	.6953	125		1/8	.9453	125
	1/4	.2031	250		1/4	.4531	250		1/4	.7031	250		1/4	.9531	250
	3/8	.2109	375		3/8	.4609	375		3/8	.7109	375		3/8	.9609	375
	1/2	.2187	500		1/2	.4687	500		1/2	.7187	500		1/2	.9687	500
	5/8	.2265	625		5/8	.4765	625		5/8	.7265	625		5/8	.9765	625
	3/4	.2343	750		3/4	.4843	750		3/4	.7343	750		3/4	.9843	750
	7/8	.2421	875		7/8	.4921	875		7/8	.7421	875		7/8	.9921	875

Friden Division, The Singer Co.

TABLE OF APPROXIMATE WEIGHTS AND MEASURES OF COMMON FOODS

Item	1 Tablespoon	1 Cup (Standard)	1 Pint		1 Quart	
	Ounces	Ounces	lb.	oz.	lb.	oz.
Allspice	1/4	4				
Apples, fresh diced					1	
Applesauce		8	1			
Bacon, diced, raw			1			
Bacon, diced, cooked					1	8
Baking Powder	1/2	6		12	1	8
Baking Soda	1/2	6	1		2	
Barley, pearl		8	1		2	
Bread Crumbs, dry		4		8	1	
Bread Crumbs, moist		2		4		8
Butter		8	1		2	
Cabbage, shredded or chopped						12
Carrots, diced, raw					1	3
Celery, chopped, raw				8	1	
Celery Seed	1/6	2⅔				
Cheese, American, ground		5		10	1	4
Cheese, shredded		4		8		
Cheese, cottage		8	1			
Chocolate, grated	1/4	4		8	1	
Chocolate, melted		8	1		2	
Cinnamon, ground	1/4	3½				
Cloves, ground	1/4	4				
Cloves, whole	1/6	2⅔				
Cocoa	1/4	3½		6½		
Coconut, shredded		2¾		5½		
Coconut, grated		2½		5		
Cornflakes						4
Cornmeal				10½	1	5
Cornstarch	1/3	5⅓				
Cracker Crumbs						10½
Cranberries, raw				8		
Currants, dried		5⅓			1	5
Curry Powder	1/4	3½				
Egg, whites (approx.)		8	1		2	
Eggs, whole (without shells)		8	1		2	
Egg, yolks		8	1		2	
Extracts (variable)	1/2					
Farina		6¼		12½	1	9
Flour, bread (sifted)		4¼		8½	1	1
Flour, cake (sifted)		3⅞		7¾		15½
Gelatin, flavored		5¾		11½	1	6
Gelatin, unflavored		5½		11	1	5
Ginger	1/2	3¾				

Item	1 Tablespoon	1 Cup (Standard)	1 Pint		1 Quart	
	Ounces	Ounces	lb.	oz.	lb.	oz.
Hominy Grits		6		12	1	8
Honey		11	1	6	2	12
Legumes:						
Beans, kidney, dry				12½	1	9
Beans, lima, dry				13	1	10
Beans, white, dry				14	1	12
Lettuce, broken or shredded						8
Mace	1/4					
Mayonnaise		8	1		2	
Milk, liquid, whole		8½	1	1	2	2
Molasses		12	1	8	3	
Mustard, ground	1/4	3¼				
Nuts, ground		4¼		9½	1	3
Nuts, pieces		4		8	1	
Nutmeg, ground	1/4	4¼				
Oats, rolled		3		6		12
Oils, cooking or salad		8		16	2	
Onions, dehydrated, flakes		2½		5		10
Onions, chopped		4		8	1	
Paprika	1/4	4				
Parsley, chopped		3		6		
Peanut Butter		9	1	2	2	4
Pepper, ground	1/4					
Peppers, green, chopped		4		8	1	
Pimientos, drained, chopped		7		14		
Potatoes, cooked, diced (approx.)				12½	1	9
Prunes, dry		5¼		10½	1	5
Raisins, seedless		5½		11	1	6
Rice, raw		8	1		2	
Sage	1/8					
Salt	1/2	8				
Seasoning, poultry	1/4					
Sugar:						
Brown (firmly packed)		7		14	1	12
Confectioners		4¾				
Granulated		8	1		2	
Syrup, corn		11	1	6	2	12
Tapioca, granules		6¼		12½	1	9
Tapioca, pearl		5¼		10½	1	6
Tea		2¾		5½		
Vinegar		8¼	1	1/2	2	1
Water		8	1		2	

DAYS EXPRESSED AS DECIMAL PART OF MONTH

	Days in Month			
Days	28	29	30	31
1	.03571	.03448	.03333	.03226
2	.07143	.06897	.06667	.06452
3	.10714	.10345	.10000	.09677
4	.14286	.13793	.13333	.12903
5	.17857	.17241	.16667	.16129
6	.21429	.20690	.20000	.19355
7	.25000	.24138	.23333	.22581
8	.28571	.27586	.26667	.25806
9	.32143	.31034	.30000	.29032
10	.35714	.34483	.33333	.32258
11	.39286	.37931	.36667	.35484
12	.42857	.41379	.40000	.38710
13	.46429	.44828	.43333	.41935
14	.50000	.48276	.46667	.45161
15	.53571	.51724	.50000	.48387
16	.57143	.55172	.53333	.51613
17	.60714	.58621	.56667	.54839
18	.64286	.62069	.60000	.58065
19	.67857	.65517	.63333	.61290
20	.71429	.68966	.66667	.64516
21	.75000	.72414	.70000	.67742
22	.78571	.75862	.73333	.70968
23	.82143	.79310	.76667	.74194
24	.85714	.82759	.80000	.77419
25	.89286	.86207	.83333	.80645
26	.92857	.89655	.86667	.83871
27	.96429	.93103	.90000	.87097
28	1.00000	.96552	.93333	.90323
29		1.00000	.96667	.93548
30			1.00000	.96774
31				1.00000

Friden Division, The Singer Co.

ABBREVIATIONS AND EQUIVALENTS

Tsp. or t. equals Teaspoon
Tbsp. or T. equals Tablespoon
Dst. equals Dessert (Soup)
  Spoon
Oz. equals Ounce
C. equals Cup
Pt. equals Pint
Qt. equals Quart

Gal. equals Gallon
Lb. or # equals Pound

Bu. equals Bushel
Cs. equals Case
Doz. equals Dozen
Ea. equals Each
Crt. equals Crate

1 Dessert (Soup) Spoon equals 2
  Teaspoons
3 Teaspoons equal 1 Tablespoon
16 Tablespoons equal 1 Cup
1 Ounce equals 1/8 Cup
2 Ounces equal 1/4 Cup
4 Ounces equal 1/2 Cup
5¼ Ounces equal 2/3 Cup
6 Ounces equal 3/4 Cup
8 Ounces equal 1 Cup

1/2 Cup equals 1 Gill
4 Gills equal 1 Pint
2 Cups equal 1 Pint
4 Cups equal 1 Quart
2 Pints equal 1 Quart
4 Quarts equal 1 Gallon
8 Quarts equal 1 Peck
4 Pecks equal 1 Bushel
1 Pound equals 16 Ounces
1 Fluid Quart equals 32 Ounces

No. 8 Ice Cream Disher equals 1/2 Cup
No. 12 Ice Cream Disher equals 1/3 Cup
No. 16 Ice Cream Disher equals 1/4 Cup

TABLE OF WEIGHTS AND THEIR APPROXIMATE EQUIVALENTS IN MEASURE

Food	Weight	Volume
Allspice	1 ounce	5 tablespoons
Apples, peeled, sliced	1 pound	1 quart
Apples, peeled, diced	1 pound	3½ cups
Apricots, dried, stewed halves	1 pound	2¼ cups
Bacon fat	6 ounces	1 cup
Bacon, raw, sliced	1 pound	15–20 slices
Bacon, raw, diced	1 pound	1 pint
Baking Powder	1 ounce	2⅔ tablespoons
Bananas, diced	1 pound	3 cups
Bananas, mashed	1 pound	2 cups
Bananas, whole	1 pound	3 medium
Barley	1 pound	2 cups
Beans, baked	1 pound	2 cups
Beans, dried, Lima, A.P.	1 pound	2½ cups
Beans, dried, Lima, Lb. A.P. after cooking	2 pounds, 9 ounces	6 cups
Beans, dried, kidney, A.P.	1 pound	2⅔ cups
Beans, dried, kidney, Lb. A.P. after cooking	2 pounds, 6 ounces	7 cups
Beans, dried, Navy, A.P.	1 pound	2½ cups
Beans, dried, Navy, Lb. A.P. after cooking	2 pounds, 3 ounces	6 cups
Beef, cooked and diced	1 pound	3 cups
Beef, raw, ground	1 pound	2 cups
Beets, medium whole	1 pound	2–3 beets
Beets, cooked and diced	1 pound	2⅓ cups
Bread, soft, broken	1 pound	10 cups
Bread, dry, broken	1 pound	9 cups
Bread crumbs, dry, ground	14 ounces	1 quart
Bread crumbs, soft	1 pound	8 cups
Butter	1 pound	2 cups
Butter	2 cups	1 pint
Cabbage, shredded, A.P.	1 pound	8 cups, lightly packed
Carrots, fresh, whole	1 pound	6 small
Carrots, diced, A.P., topped	1 pound	3¼ cups
Carrots, ground, raw	1 pound	3¼ cups
Carrots, diced, cooked	1 pound	3 cups
Carrots, raw, sliced	1 pound	3 cups
Catsup	1 pound	2 cups
Celery, diced	1 pound	1 quart
Celery Salt	1 ounce	2 tablespoons
Celery Seed	1 ounce	4 tablespoons
Cheese, cottage	1 pound	2 cups
Cheese, grated or ground	1 pound	1 quart
Cherries, Maraschino, whole or chopped	1 pound	2 cups
Chicken, diced, cooked	1 pound	3 cups
Chocolate, grated	4 ounces	1 cup
Chocolate, melted	8 ounces	1 cup
Cinnamon, ground	1 ounce	4 tablespoons

Food	Weight	Volume
Cloves, ground	1 ounce	5 tablespoons
Cloves, whole	3 ounces	1 cup
Cocoa	4 ounces	1 cup
Coconut, medium shredded	1 pound	6 cups
Coffee	1 pound	5 cups
Corn Syrup	10 ounces	1 cup
Corn, whole kernel	1 pound	2 cups
Corned Beef, canned	1 pound	2½ cups
Cornflakes	1 pound	4 quarts
Cornmeal	5½ ounces	1 cup
Cornstarch	5 ounces	1 cup
Cracker crumbs	1 pound	6 cups
Cranberries	1 pound	1 quart
Cranberry, pulp	1 pound	2 cups
Cream of Tartar	1 ounce	3 tablespoons
Cucumbers, diced	1 pound	3½ cups
Cucumbers, 50-60 1/8" slices	1 pound	2-3 large
Curry Powder	1 ounce	4 tablespoons
Dates, chopped	1 pound	3 cups
Dates, pitted	1 pound	3 cups
Dates, whole	1 pound	2½ cups
Eggplant, diced	1 pound	3 cups
Eggs, hard-cooked, chopped	1 pound	3 cups
Eggs, uncooked, whole	1¾ ounces	1 egg
Eggs, uncooked, whole	1 pound	9 eggs
Eggs, uncooked, whole	1 pound	2 cups
Eggs, uncooked, whites	1 ounce	1 egg white
Eggs, uncooked, whites	1 pound	15 egg whites
Eggs, uncooked, whites	1 pound	2 cups
Eggs, uncooked, yolks	3/4 ounce	1 yolk
Eggs, uncooked, yolks	1 pound	12 yolks
Eggs, uncooked, yolks	1 pound, 2 ounces	2 cups
Farina, raw	6 ounces	1 cup
Flour, sifted once before measuring		
Bread	1 pound	4-4½ cups
Cake, Pastry and Rye	1 pound	4-5 cups
Whole Wheat	1 pound	3-4 cups
Gelatin, flavored	1 pound	2 cups
Gelatin, granulated	4 ounces	1 cup
Gelatin, granulated	1 ounce	4 tablespoons
Ginger, ground	1 ounce	4 tablespoons
Grapes, purple	1 pound	2¼ cups
Grapes, white, seedless	1 pound	3 cups
Green peppers	1 pound	6 medium
Green peppers, chopped	1 pound	1 quart
Ham, cooked, diced	1 pound	3 cups

Food	Weight	Volume
Ham, ground	1 pound	2 cups
Honey	12 ounces	1 cup
Horseradish, prepared	8 ounces	1 cup
Ice Cream	6 pounds	1 gallon
Jam	1 pound	1⅓ cups
Jelly	1 pound	1½ cups
Lard	8 ounces	1 cup
Lemon Juice	8 ounces (4-5 lemons)	1 cup
Lemon rind, grated	1 ounce	3 tablespoons
Lemons, size 300	1 pound	4 lemons
Lettuce, shredded	8 ounces	1 quart
Lettuce, medium heads	1 pound	2 heads
Macaroni, cooked	1 pound	2½ cups
Macaroni, cut, A.P.	18 ounces	1 quart
Macaroni, 1 lb. before cooking, after cooking	5 pounds	2¼ quarts
Macaroni, cut, A.P.	4½ ounces	1 cup
Mayonnaise	1 pound	2 cups
Meat, cooked, diced	1 pound	2½ cups
Milk, fresh	8 ounces	1 cup
Milk, evaporated	1 pound	2 cups
Milk, evaporated, No. 1 tall can	14½ ounces	1⅔ cups
Molasses	1 pound	1⅓ cups
Molasses	11 ounces	1 cup
Mushrooms, canned	1 pound	2 cups
Mushrooms, fresh, sliced	1 pound	1¾ quarts
Mustard, dry	1 ounce	4 tablespoons
Mustard Seed	1 ounce	3 tablespoons
Mustard, prepared	1/2 ounce	1 tablespoon
Mustard, prepared	10 ounces	1 cup
Noodles, cooked	6 pounds	1 gallon
Noodles, cooked	1 pound	2¾ cups
Noodles, raw, dry	12 ounces	1 quart
Nutmeg, ground	1 ounce	4 tablespoons
Nuts, chopped	4 ounces	1 cup
Oatmeal, raw	1 pound	5 cups
Oil	1 pound	2⅛ cups
Olives, chopped	6 ounces	1 cup
Onions, chopped	1 pound	3 cups
Onions, grated	1 ounce	1⅓ tablespoons
Onions, sliced	1 pound	4 cups
Onions, dehydrated	2 ounces	1 cup
Onions, dehydrated, reconstituted	2 ounces 1 quart water	1 pound, chopped
Oranges, 150 size	1 pound	2 each
Oranges	8 medium	1 quart sections
Oranges	3 medium	1 cup juice
Oranges, grated rind	3 ounces	1 cup
Oysters	3/4 ounce	1 large
Oysters	1 quart	60-100 small
Oysters	1 quart	24-40 large
Oysters	1 pound	2¼ cups
Paprika	1 ounce	4 tablespoons
Parsley, chopped	1 ounce	3/4 cup
Vinegar	8 ounces	1 cup
Water	8 ounces	1 cup
Wheat Cereal, granulated	1 pound	1½ pints
White Sauce	9 ounces	1 cup
Worcestershire Sauce	2⅓ ounces	1⅓ tablespoons
Yeast, compressed	1/2 ounce	1 cake

SIZE PORTIONS AND SERVING UTENSILS REQUIRED

Food	Size Portions	Serving Utensil
Beverages		
Cocoa	8 oz.	
Coffee	8 oz.	
Milk	8 oz.	
Tea	8 oz.	
Breads		
White, Rye, Whole Wheat, French or Italian	2 slices	Tongs
Biscuits	2 each	Tongs
Cornbread	1 piece -2½" × 2½" × 2	Spatula
Crackers	4 each	Tongs
French Toast	1½ slices	Spatula
Wheat Cakes	3 cakes	Spatula
Rolls	2 each	Tongs
Butter or Margarine	1 pat	Tongs or Fork
Cereals		
Cracked Wheat	6 oz.	6 oz. ladle
Farina or Cream of Wheat	6 oz.	6 oz. ladle
Oatmeal	6 oz.	6 oz. ladle
Steamed Rice	6 oz.	6 oz. ladle
Whole Wheat	6 oz.	6 oz. ladle
Bran Flakes	1 oz. or 1 pkg. (indiv.)	6 oz. ladle
Corn Flakes	1 oz. or 1 pkg. (indiv.)	6 oz. ladle
"K" Cereal	1 oz. or 1 pkg. (indiv.)	6 oz. ladle
Raisin Bran Flakes	1 oz. or 1 pkg. (indiv.)	6 oz. ladle
Rice Krispies	1 oz. or 1 pkg. (indiv.)	6 oz. ladle
Sugar Corn Pops	1 oz. or 1 pkg. (indiv.)	6 oz. ladle
Sugar Frosted Flakes	1 oz. or 1 pkg. (indiv.)	6 oz. ladle
Sugar Krisp	1 oz. or 1 pkg. (indiv.)	6 oz. ladle
Assorted Individuals	1 pkg.	
Cheese		
Cottage Cheese	3 oz.	#12 scoop
Sliced Cheese	2 oz.	Tongs or Spatula
Eggs		
Whole	6 servings per dz. boiled	Tongs
	8 servings per dz. fried	Spatula
Scrambled	3 oz.	Serving Spoon
Fruits and Juices		
Apples, eating	1 each	
Apples, Baked	1 each	Serving spoon
Banana	1 each	
Orange	1 each	
Grapefruit	1/2 grapefruit	
Applesauce	4 oz.	4 oz. ladle
Apricots	4 oz.	4 oz. ladle
Cherries	4 oz.	4 oz. ladle

Food	Size Portions	Serving Utensil
Fruit Cocktail	4 oz.	4 oz. ladle
Peach Halves	4 oz.	4 oz. ladle
Peaches Sliced	4 oz.	4 oz. ladle
Pear Halves	4 oz.	4 oz. ladle
Pineapple, sliced	4 oz.	4 oz. ladle
Pineapple, crushed	4 oz.	4 oz. ladle
Prune Plums	4 oz.	4 oz. ladle
Strawberries	4 oz.	4 oz. ladle
Stewed Apricots	4 oz.	4 oz. ladle
Stewed Peaches	4 oz.	4 oz. ladle
Stewed Prunes	4 oz.	4 oz. ladle
Stewed Mixed Fruit Nuggets	4 oz.	4 oz. ladle
Grapefruit Juice	4 oz.	4 oz. ladle
Orange Juice	4 oz.	4 oz. ladle
Pineapple Juice	4 oz.	4 oz. ladle
Tomato Juice	4 oz.	4 oz. ladle
Jelly, Jam, Honey, Peanut Butter	1 oz.	1 oz. ladle
Meat		
Baked Beef Loaf	1-3 oz. slice	Spatula
Beef ala Mode	4 oz.	Serving spoon
Beef Barbecue, Swiss Style	4 oz.	Serving spoon
Beef Liver	2½ oz.	Tongs
Beef Liver Loaf	1-3 oz. slice	Spatula
Beef and Noodles	8 oz.	8 oz. ladle
Beef Stew	8 oz.	8 oz. ladle
Boiled Beef (New England)	3 oz.	Tongs
Braised Beef Cubes & Noodles	8 oz.	8 oz. ladle
Braised Beef Tenderloin Tip	4 oz.	Serving spoon
Chili Con Carne	8 oz.	8 oz. ladle
Chop Suey	6 oz.	6 oz. ladle
Corned Beef Hash	6 oz.	6 oz. ladle
Creamed Chipped Beef	6 oz.	6 oz. ladle
Hamburger Patties	2-2 oz. or 1-4 oz.	Tongs or spatula
Italian Spaghetti	8 oz.	8 oz. ladle
Pot Roast of Beef	3 oz.	Tongs or fork
Roast Round of Beef	3 oz.	Tongs or fork
Round Steak, oven-fried	3 oz.	Tongs or fork
Salisbury Steak	3 oz.	Tongs or spatula
Swiss Steak	3 oz. - 1 piece	Tongs or fork
Baked Lamb Loaf	3 oz. - 1 slice	Spatula
Roast Leg O' Lamb	3 oz.	Tongs or fork
Canadian Bacon	2 slices (cut 14 per lb.)	Tongs or fork
Baked Ham	3 oz.	Tongs or fork
Fried Sausage Links	2 each	Tongs
Smoked Pork Butts	2-1½ oz. slices	Tongs or fork
Pork Chops, Baked	1 each (cut 3 per lb.)	Tongs or fork
Pork Chops, Breaded	2 each (cut 6 per pound)	Tongs or fork
Pork Roast	3 oz.	Tongs or fork
Pork Steaks	1 steak - 3½ oz.	Tongs or fork
Spareribs	8 oz.	Tongs or fork
Veal Loaf	3 oz. - 1 slice	Spatula
Roast Leg O' Veal	3 oz.	Tongs or fork
Veal cutlets, Breaded	1 cutlet (cut 5 per pound)	Tongs or fork
Cold Cuts	2 oz.	Tongs or fork
Poultry		
Baked Chicken	3 oz.	Spatula or serving spoon

Food	Size Portions	Serving Utensil
Chicken ala King	6 oz.	6 oz. ladle
Creamed Chicken	6 oz.	6 oz. ladle
Fricassee of Chicken & Noodles	8 oz.	8 oz. ladle
Chicken Pie	6 oz.	Serving spoon
Fried Chicken	1/4 or 6 oz.	Tongs
Fish		
Baked Haddock	4 oz.	Tongs or spatula
Baked Halibut	4 oz.	Tongs or spatula
Fried Perch	4 oz.	Tongs or spatula
Creamed Salmon	6 oz.	6 oz. ladle
Salmon Loaf	3 oz.	Spatula
Salmon Salad	4 oz.	No. 8 scoop (level)
Tunafish ala King	6 oz.	6 oz. ladle
Tunafish Salad	4 oz.	No. 8 scoop (level)
Tunafish & Macaroni Casserole	6 oz.	Serving spoon
Tunafish & Macaroni Salad	4 oz.	No. 8 scoop (level)
Tunafish & Noodle Casserole	6 oz.	Serving spoon
Fish Croquettes	2 each (#12 scoop)	Tongs or spatula
Potatoes		
Au Gratin Potatoes	4 oz.	Serving spoon
Baked Potato	1 med. (6 oz. with skin)	Tongs
Creamed Potatoes	4 oz.	Serving spoon
Escalloped Potatoes	4 oz.	Serving spoon
Hot Potato Salad	4 oz.	Serving spoon
Mashed Potatoes	4 oz.	No. 8 scoop (level)
Parsley Potatoes	4 oz. (2 small)	Serving spoon
Potatoes in Cheese Sauce	4 oz.	Serving spoon
Sweet Potatoes, Candied	4 oz.	Serving spoon
Sweet Potatoes, Glazed	4 oz.	Serving spoon
Sweet Potatoes, Mashed	4 oz.	Serving spoon
Sweet Potato, Baked	1 each	Serving spoon
Potato Substitutes		
Baked Beans	4 oz.	Serving spoon
Boiled Navy Beans	4 oz.	Serving spoon
Spanish Pinto Beans	4 oz.	Serving spoon
Baked Macaroni and Cheese	4 oz.	Serving spoon
Buttered Noodles	4 oz.	Serving spoon
Corn Bread Dressing	4 oz.	Serving spoon
Bread Dressing	4 oz.	Spatula or serving spoon
Salads		
Beef and Cabbage Salad	3 oz.	Serving spoon
Cabbage-Apple Salad	3 oz.	Serving spoon
Carrot-Apple-Raisin Salad	3 oz.	Serving spoon
Chef's Salad	3 oz.	Serving spoon
Cole Slaw	3 oz.	Serving spoon
Cranberry-Orange Relish	3 oz.	Serving spoon
Cucumber-Onion Salad	3 oz.	Serving spoon
Kidney Bean Salad	4 oz.	No. 8 scoop (level)
Mixed Vegetable Salad	3 oz.	Serving spoon
Molded Cherry Salad	4 oz.	Spatula
Pear - Lime Gelatin Salad	4 oz.	Spatula
Perfection Salad	4 oz.	Spatula
Tomato Aspic	4 oz.	Spatula
Tossed Salad	3 oz.	Serving spoon
Waldorf Salad	3 oz.	Serving spoon
Sliced Tomatoes	3 oz. (2 slices)	Tongs or fork

Food	Size Portions	Serving Utensil
Salad Dressings		
All Dressings	1 oz.	1 oz. ladle
Sauces and Gravies	1-2 oz.	1 or 2 oz. ladle
Soups		
All Soups and Oyster Stew	8 oz.	8 oz. ladle
Vegetables		
All Vegetables	4 oz.	Serving spoon
Desserts		
Cakes, Cottage Pudding	1 piece - 2½″ × 2½″ × 2″	Spatula
Pies	1 piece (cut 6 per 9″ pie)	Spatula
Cobblers	1 piece - 2½″ × 2½″ × 2″	Spatula
Cookies	2 each	Tongs
Jello	4 oz.	Spatula
Custard	4 oz.	Spatula or 4 oz. ladle
Puddings	4 oz.	4 oz. ladle
Ice Cream	24 servings per gallon	No. 12 scoop

PORTION CONTROL DATA

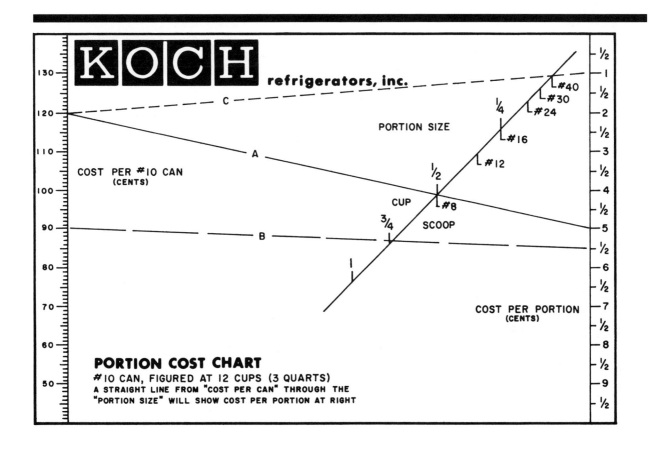

Above is an example of the KOCH-designed work saver charts which are useful in food cost control.

This chart, called a nomograph, is valuable to the foodservice operator because it reduces the time spent on portion data and reduces mathematical errors.

It is made up of three scales:

1. Cost per unit of purchase.

2. Portion size.

3. Cost per portion.

HOW TO USE

To use Portion Cost Charts, factors on two of the three scales must be known. Place a straight-edge through the two known factors. The third factor is located at the point where the straight-edge crosses the third scale. Examples, using the above chart:

Example No. 1 (Line A). To determine the portion cost of an item served in 1/2-cup portions and which cost $1.20 per #10 can:

Place a straight-edge through the "120" on the left-hand scale and "1/2-cup" on the center scale. The point where it crosses the right-hand scale is "5." This shows a portion cost of 5¢ per serving.

Example No. 2 (Line B). To determine the portion size of an item that costs 90¢ for a #10 can and which is budgeted at 5½¢ a portion:

Place the straight-edge through "90" on the left-hand scale and "5½" on the right-hand scale. The point where it crosses the center scale is "3/4." This means that 3/4-cup can be served to fit the budget allowance of 5½¢ per serving.

Example No. 3 (Line C). To decide which relish to use if the budget allows 1¢ for a #40 scoop:

Place the straight-edge through "#40" on the center scale and "1" on the right-hand scale. It will cross the left-hand scale at "120," which means that an appropriate relish that costs about $1.20 for a #10 can may be used.

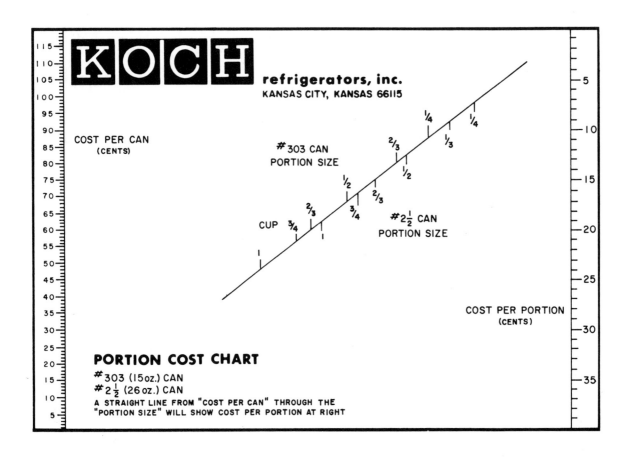

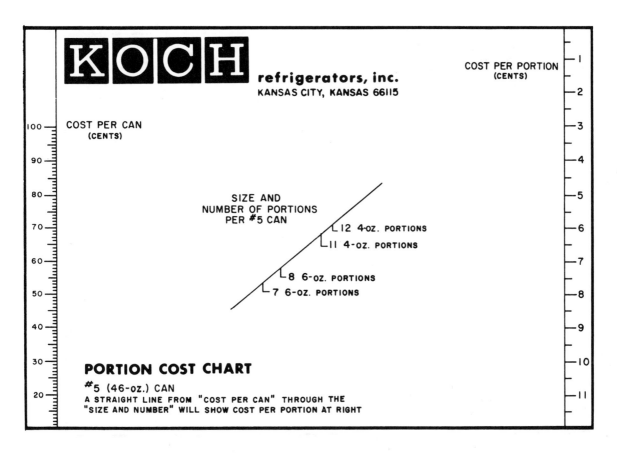

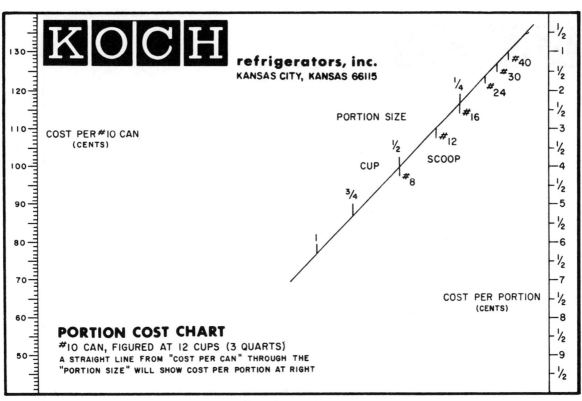

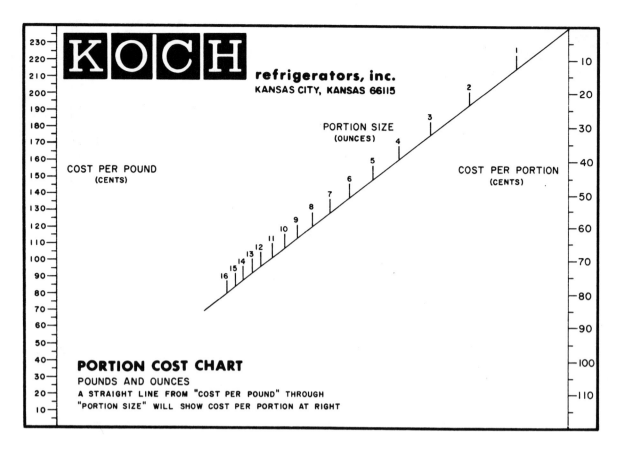
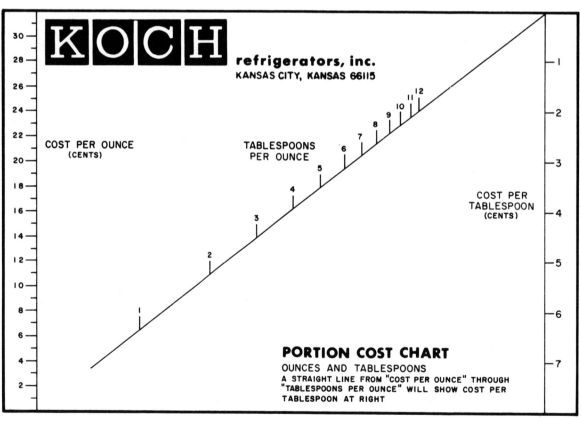

TABLESPOONS PER OUNCE OF COMMONLY USED FOODS

Item	Tbsp Per Oz.	Cost Per Oz.	Date	Item	Tbsp Per Oz.	Cost Per Oz.	Date	Item	Tbsp Per Oz.	Cost Per Oz.	Date
Allspice, Ground	5			Curry Powder	8			Pickling Spice	8		
Allspice, Whole	10			Flour, All Purpose	4			Poppy Seeds	6		
Baking Powder	3			Garlic Salt	3			Poultry Seasoning	10		
Basil	8			Ginger, Ground	8			Sage, Ground	8		
Cassia Buds	6			Honey	1			Salt	2		
Celery Salt	4			Mace	6			Shortening	2		
Celery Seed	6			Margarine	2			Soda, Baking	3		
Chili Powder	4			Marjoram	10			Thyme	12		
Cinnamon, Ground	6			Milk, Dry	4			Vanilla Extract	2		
Cloves, Ground	5			Mustard, Dry	5						
Cloves, Whole	6			Mustard, Prepared	4						
Cocoa	4			Nutmeg, Ground	6						
Coconut, Grated	6			Onion Juice	2						
Coffee	5			Onion Salt	3						
Coriander	10			Paprika	6						
Cornmeal	3			Pepper, Ground Black	5						
Cornstarch	3			Pepper, Red	7						
Cream of Tartar	3			Pepper, White	6						

COMMON CONTAINER SIZES

The labels of cans or jars of identical size may show a net weight for one product that differs slightly from the net weight on the label of another product, due to the difference in the density of the food. An example would be pork and beans (1 lb.), blueberries (14 oz.), in the same size can.

Industry Term	Container Consumer Description — Approx. Net Weight or Fluid Measure (Check Label)	Approx. Cups	Principal Products
8 oz.	8 oz.	1	Fruits, vegetables, *specialties for small families. 2 servings.
Picnic	10½ to 12 oz.	1¼	Mainly condensed soups. Some fruits, vegetables, meat, fish, *specialties. 2 to 3 servings.
12 oz. (vac.)	12 oz.	1½	Principally for vacuum pack corn. 3 to 4 servings.
No. 300	14 to 16 oz. (14 oz. to 1 lb.)	1¾	Pork and beans, baked beans, meat products, cranberry sauce, blueberries, *specialties. 3 to 4 servings.
No. 303	16 to 17 oz. (1 lb. to 1 lb. 1 oz.)	2	Principal size for fruits and vegetables. Some meat products, ready-to-serve soups, *specialties. 4 servings.
No. 2	20 oz. 18 fl. oz. (1 lb. 4 oz.) (1 pt. 2 fl. oz.)	2½	**Juices, ready-to-serve soups, some *specialties, pineapple, apple slices. No longer in popular use for most fruits and vegetables. 5 servings.
No. 2½	27 to 29 oz. (1 lb. 11 oz. to 1 lb. 13 oz.)	3½	Fruit, some vegetables (pumpkin, sauerkraut, spinach and other greens, tomatoes). 5 to 7 servings.
No. 3 cyl. or 46 fl. oz.	51 oz. 46 fl. oz. (3 lb. 3 oz.) 1 qt. 14 fl. oz.	5¾	Fruit and vegetable juices**, pork and beans. Institutional size for condensed soups, some vegetables. 10 to 12 servings.
No. 10	6½ lb. to 7 lb. 5 oz.	12-13	Institutional size for fruits, vegetables, and some other foods. 25 servings.

*SPECIALTIES: Usually a food combination such as macaroni, spaghetti, Spanish style rice, Mexican type foods, Chinese foods, tomato aspic, etc.
**Juices are now being packed in a number of can sizes.

Strained and homogenized foods for infants, and chopped junior foods, come in small jars and cans suitable for the smaller servings used. The weight is given on the label.

Meats, poultry, fish, and seafood are almost entirely advertised and sold under weight terminology.

SUBSTITUTING ONE CAN FOR ANOTHER SIZE—FOR INSTITUTIONAL USE

		Approx.	
1 No. 10 can equals	7 No. 303	(1 lb.)	cans
1 No. 10 can equals	5 No. 2	(1 lb. 4 oz.)	cans
1 No. 10 can equals	4 No. 2½	(1 lb. 13 oz.)	cans
1 No. 10 can equals	2 No. 3 Cyl.	(46 to 50 oz.)	cans

A GUIDE TO COMMON CAN SIZES

No. 1/4 Flat Can	4¾ oz. approximately 1/2 cup
No. 1/2 Flat Can	7¾ oz.-8½ oz. approximately 1 cup
No. 1 Tall Can	12 to 16 oz. approximately 2 cups
No. 2 Can	1 lb. 2 oz. to 1 lb. 8 oz. approximately 2½ cups
No. 2½ Can	1 lb. 10 oz. to 2 lb. 3 oz. approximately 3½ cups
No. 5 Can	3 lb. 9 oz. approximately 6 cups
No. 10 Can	6 lb. to 8 lb. approximately 13 cups

Citing American Can Company, several publications.
Home Economics—Consumer Services National Canners Association 1133 20th St. N.W., Washington, D.C. 20036

FROZEN FOOD CAPACITY IN 12½ CUBIC FOOT DOOR OPENING

	Boxes and/or Cartons			Number Boxes and/or Cartons Stored per	
ITEM	Capacity	Dimension in Inches	Cubic Feet	Level	Door Opening (Half-Height)
VEGETABLE	2½ lbs.	9⅝ × 5⅜ × 2½	13	25	125
VEGETABLE	12 oz.	5⅜ × 4 × 1½	192	113	565
POTATO, French Fry Cut	30 lbs.	18 × 11½ × 10⅛	—	2	4
FISH					
Breaded Fish Stick	6 lbs.	10⅛ × 8 × 2¾	7	22	66
Lobster Tail	5 lbs.	14⅞ × 7¼ × 3⅜	4	11	36
Rainbow Trout	5 lbs.	12⅞ × 8¼ × 2¾	6	17	51
Shrimp	5 lbs.	11⅝ × 6¼ × 2¾	8	17	74
Shrimp	2 lbs.	7⅝ × 5⅜ × 2⅛	19	56	180
POULTRY					
Chicken Parts	10 lbs.	18¼ × 10¾ × 2¾	3	10	26
Turkey Roll	11 lbs.	17⅛ × 5 × 5	4	6	30
MEAT, Ground Beef	50 lbs.	20¾ × 15¾ × 5¼	1	1	8
DAIRY PRODUCTS					
Butter	32 lbs.	11 × 11 × 11	1	4	8
Cheese	30 lbs.	12 × 12 × 8½	1	3	9
PIE	1 lb. 8 oz.	8¼ × 8¼ × 1⅝	16	7	112
ICE CREAM					
Rectangular	1 pt.	3⅜ × 4 × 2½	55	63	441
Rectangular	1/2 gal.	6⅞ × 4¾ × 3½	15	26	130
Round	2½ gal.	9½D × 10H	1	7	14
CANS AND/OR JARS			Number Cans or Jars Stored per		
ITEM	Capacity	Dimension in Inches	Cubic Feet	Level	Door Opening (Half-Height)
FRUIT	30 lbs.	10D; 13H	1	6	12
	10 lbs.	7¼D; 8¾H	3	11	33
	8½ lb.	6¼D; 8¾H	5	14	52
	#10 Can	6⅛D; 7H	10	16	48
	6½ lb.	6⅛D; 7H	6	16	48
	4½ lb.	6⅛D; 4⅝H	10	16	80
	#5 Can	4¼D; 7H	13	28	84
FRUIT JUICE					
Orange Juice	2½ lbs.	4⅛D; 5½H	19	28	112
Concentrated Orange	12 oz.	2¾D; 4⅞H	49	86	430
Concentrated Orange	32 oz.	4D; 5⅝H	19	39	156
Concentrated Lemon	18 oz.	3½D; 4⅝H	34	50	250
HORSERADISH	#303 Can	3⅛D; 4¾H	37	56	280

Koch Refrigerators, Inc.

ENERGY COST OF DOING WORK*

Activity	Calories Required Per Hour by Average Man
Sleeping	65
Sitting at rest	100
Standing relaxed	105
Standing at attention	115
Singing	122
Dishwashing	144
Sweeping bare floor (38 strokes per minute)	169
Walking slowly (2.6 miles per hour)	200
Walking moderately (3.75 miles per hour)	300
Walking down stairs	364
Severe exercise	450
Walking very fast (5.3 miles per hour)	650
Walking up stairs	1100

*Taken from p. 185 *Chemistry of Foods and Nutrition* by C. Sherman. MacMillan Company, 1941.

COMMON FOOD PORTIONS

Equivalents by Weight

1 pound (16 ounces)	453.6 grams
1 ounce	28.35 grams
3½ ounces	100 grams

Equivalents by Volume (All Measurements Level)

1 quart	4 cups
1 cup	8 fluid ounces
	1/2 pint
	16 tablespoons
2 tablespoons	1 fluid ounce
1 tablespoon	3 teaspoons
1 pound butter or margarine	4 sticks
	2 cups
	64 pats or squares
1 stick butter or margarine	1/2 cup (approx.)
	16 pats or squares

Citing Helen Belinkie, *The Gourmet in the Low-Calorie Kitchen*, David McKay Company, Inc., New York, 1961.

SUGGESTED COOKING GUIDE FOR GRIDDLE ITEMS

Food	Control Set	Time in Min.
Hamburgers	350°F	3–4
Cheeseburgers	350°F	3–4
Cheese Sandwich	375°F	3–4
Ham Salad Sandwich	375°F	3–4
Frankfurters	325°F	2–3
Minute Steak—medium	400°F	3–4
Club Steaks—inch thick, med.	400°F	3–5
Ham Steaks	375°F	3–4
Beef Tenderloin	400°F	3–4
Boiled Ham	375°F	2
Corned Beef Patties	350°F	2–3
Bacon	350°F	2–3
Canadian Bacon	350°F	2–3
Sausage Links	350°F	3
Sausage Patties	350°F	3
French Toast	350°F	2–3
Pancakes	375°F	2
American Fries	375°F	3–4
Potato Patties	375°F	3–4
Scrambled Eggs	300°F	1–2
Hard Fried Eggs	300°F	3
Soft Fried Eggs	300°F	2
Sunny Side Up Eggs	300°F	2

WEIGHT CONVERSION CHARTS*

These two conversion charts, the OUNCE CHART below, and the POUND CHART that follows, have been designed to permit easy adjustment of basic recipes for the number of portions actually needed.

Example: A basic 100-portion recipe calls for 7 ounces of a particular ingredient. When adjusting to 25 portions, find the column headed "100 Portions" and move down to the space marked "7 oz." Then move across this space, horizontally to the left, to the column headed "25" Portions. The figure "1¾" then appears as the number of ounces of the ingredient needed.

OUNCE CHART

25 Portions	50 Portions	75 Portions	100 Portions	200 Portions	300 Portions	400 Portions	500 Portions	600 Portions	700 Portions	800 Portions	900 Portions	1000 Portions
—	—	1/8 oz	1/8 oz	1/4 oz	3/8 oz	1/2 oz	5/8 oz	3/4 oz	7/8 oz	1 oz	1⅛ oz	1¼ oz
—	—	1/4 oz	1/4 oz	1/3 oz	1/3 oz	2/3 oz	3/4 oz	1 oz	1¼ oz	1⅓ oz	1½ oz	1⅔ oz
—	—	1/5 oz	1/5 oz	2/3 oz	3/5 oz	4/5 oz	1 oz	1⅕ oz	1⅖ oz	1⅗ oz	1⅘ oz	2 oz
—	—	1/4 oz	1/4 oz	1/2 oz	3/4 oz	1 oz	1¼ oz	1½ oz	1¾ oz	2 oz	2¼ oz	2½ oz
—	—	1/3 oz	1/3 oz	2/3 oz	1 oz	1⅓ oz	1⅔ oz	2 oz	2⅓ oz	2⅓ oz	3 oz	3⅓ oz
—	1/4 oz	3/8 oz	1/3 oz	1 oz	1½ oz	2 oz	2½ oz	3 oz	3½ oz	4 oz	4½ oz	5 oz
—	1/3 oz	1/2 oz	2/3 oz	1⅓ oz	2 oz	2⅔ oz	3⅓ oz	4 oz	4⅔ oz	5⅓ oz	6 oz	6⅔ oz
—	3/8 oz	5/8 oz	3/4 oz	1½ oz	2¼ oz	3 oz	3¾ oz	4½ oz	5¼ oz	6 oz	6¾ oz	7½ oz
1/4 oz	1/2 oz	3/4 oz	1 oz	2 oz	3 oz	4 oz	5 oz	6 oz	7 oz	8 oz	9 oz	10 oz
1/3 oz	1 oz	1½ oz	2 oz	4 oz	6 oz	8 oz	10 oz	12 oz	14 oz	1#	1#2 oz	1#4 oz
3/4 oz	1½ oz	2¼ oz	3 oz	6 oz	9 oz	12 oz	15 oz	1#2 oz	1#5 oz	1#8 oz	1#11oz	1#14oz
1 oz	2 oz	3 oz	4 oz	8 oz	12 oz	1#	1#4 oz	1#8 oz	1#12oz	2#	2#4 oz	2#8 oz
1¼ oz	2½ oz	3¾ oz	5 oz	10 oz	15 oz	1#4 oz	1#9 oz	1#14oz	2#3 oz	2#8 oz	2#13 oz	3#2 oz
1½ oz	3 oz	4½ oz	6 oz	12 oz	1#2 oz	1#8 oz	1#14oz	2#4 oz	2#10 oz	3#	3#6 oz	3#12oz
1¾ oz	3½ oz	5¼ oz	7 oz	14 oz	1#5 oz	1#12oz	2#3 oz	2#10 oz	3#1 oz	3#8 oz	3#15oz	4#6 oz
2 oz	4 oz	6 oz	8 oz	1#	1#8 oz	2#	2#8 oz	3#	3#8 oz	4#	4#8 oz	5#
2¼ oz	4½ oz	6¾ oz	9 oz	1#2 oz	1#11oz	2#4 oz	2#13oz	3#6 oz	3#15oz	4#8 oz	5#1 oz	5#10oz
2½ oz	5 oz	7½ oz	10 oz	1#4 oz	1#14oz	2#8 oz	3#2 oz	3#12oz	4#6 oz	5#	5#10oz	6#4 oz
2¾ oz	5½ oz	8¼ oz	11 oz	1#6 oz	2#1 oz	2#12oz	3#7 oz	4#2 oz	4#13oz	5#8 oz	6#3 oz	6#14 oz
3 oz	6 oz	9 oz	12 oz	1#8 oz	2#4 oz	3#	3#12oz	4#8 oz	5#4 oz	6#	6#12oz	7#8 oz
3¼ oz	6½ oz	9¾ oz	13 oz	1#10 oz	2#7 oz	3#4 oz	4#1 oz	4#14oz	5#11oz	6#8 oz	7#5 oz	8#2 oz
3½ oz	7 oz	10½ oz	14 oz	1#12 oz	2#10 oz	3#8 oz	4#6 oz	5#4 oz	6#2 oz	7#	7#14oz	8#12oz
3¾ oz	7½ oz	11¼ oz	15 oz	1#14 oz	2#13oz	3#12oz	4#11 oz	5#10oz	6#9 oz	7#8 oz	8#7 oz	9#6 oz

POUND CHART

25 Portions	50 Portions	75 Portions	100 Portions	200 Portions	300 Portions	400 Portions	500 Portions	600 Portions	700 Portions	800 Portions	900 Portions	1000 Portions
4 oz.	8 oz.	12 oz.	1#	2#	3#	4#	5#	6#	7#	8#	9#	10#
5 oz.	10 oz.	15 oz.	1#4 oz.	2#8 oz.	3#12oz.	5#	6#4 oz.	7#8 oz.	8#12oz.	10#	11#4 oz.	12#8 oz.
6 oz.	12 oz.	1#2 oz.	1#8 oz.	3#	4#8 oz.	6#	7#8 oz.	9#	10#8oz.	12#	13#8oz.	15#
7 oz.	14 oz.	1#5 oz.	1#12oz.	3#8 oz.	5#4 oz.	7#	8#12oz.	10#8oz.	12#4oz.	14#	15#12oz.	17#8oz.
8 oz.	1#	1#8 oz.	2#	4#	6#	8#	10#	12#	14#	16#	18#	20#
9 oz.	1#2 oz.	1#11oz.	2#4 oz.	4#8 oz.	6#12oz.	9#	11#4oz.	13#8oz.	15#12oz.	18#	20#4oz.	22#8oz.
10 oz.	1#4 oz.	1#14oz.	2#8 oz.	5#	7#8 oz.	10#	12#8oz.	15#	17#8oz.	20#	22#8oz.	25#
11 oz.	1#6 oz.	2#1 oz.	2#12oz.	5#8 oz.	8#4 oz.	11#	13#12oz.	16#8oz.	19#4oz.	22#	24#12oz.	27#8oz.
12 oz.	1#8 oz.	2#4 oz.	3#	6#	9#	12#	15#	18#	21#	24#	27#	30#
13 oz.	1#10oz.	2#7 oz.	3#4 oz.	6#8 oz.	9#12oz.	13#	16#4oz.	19#8oz.	22#12oz.	26#	29#4oz.	32#8oz.
14 oz.	1#12oz.	2#10oz.	3#8 oz.	7#	10#8oz.	14#	17#8oz.	21#	24#8oz.	28#	31#8oz.	35#
15 oz.	1#14oz.	2#13oz.	3#12oz.	7#8oz.	11#4oz.	15#	18#12oz.	22#8oz.	26#4oz.	30#	33#12oz.	37#8oz.
1#	2#	3#	4#	8#	12#	16#	20#	24#	28#	32#	36#	40#
1#1 oz.	2#2 oz.	3#3 oz.	4#4 oz.	8#8 oz.	12#12oz.	17#	21#4oz.	25#8oz.	29#12oz.	34#	38#4oz.	42#8oz.
1#2 oz.	2#4 oz.	3#6oz.	4#8 oz.	9#	13#8oz.	18#	22#8oz.	27#	31#8oz.	36#	40#8oz.	45#
1#3 oz.	2#6 oz.	3#9 oz.	4#12oz.	9#8 oz.	14#4oz.	19#	23#12oz.	28#8oz.	33#4oz.	38#	42#12oz.	47#8oz.
1#4 oz.	2#8 oz.	3#12oz.	5#	10#	15#	20#	25#	30#	35#	40#	45#	50#
1#5 oz.	2#10oz.	3#15oz.	5#4oz.	10#8oz.	15#12oz.	21#	26#4oz.	31#8oz.	36#12oz.	42#	47#4oz.	52#8oz.
1#6 oz.	2#12oz.	4#2 oz.	5#8 oz.	11#	16#8oz.	22#	27#8oz.	33#	38#8oz.	44#	49#8oz.	55#
1#7 oz.	2#14oz.	4#5oz.	5#12oz.	11#8oz.	17#4oz.	23#	28#12oz.	34#8oz.	40#4oz.	46#	51#12oz.	57#8oz.
1#8 oz.	3#	4#8oz.	6#	12#	18#	24#	30#	36#	42#	48#	54#	60#
1#12oz.	3#8oz.	5#4oz.	7#	14#	21#	28#	35#	42#	49#	56#	63#	70#
2#	4#	6#	8#	16#	24#	32#	40#	48#	56#	64#	72#	80#
2#4oz.	4#8oz.	6#12oz.	9#	18#	27#	36#	45#	54#	63#	72#	81#	90#
2#8oz.	5#	7#8oz.	10#	20#	30#	40#	50#	60#	70#	80#	90#	100#
2#12oz.	5#8oz.	8#4oz.	11#	22#	33#	44#	55#	66#	77#	88#	99#	110#
3#	6#	9#	12#	24#	36#	48#	60#	72#	84#	96#	108#	120#
3#12oz.	7#8oz.	11#4oz.	15#	30#	45#	60#	75#	90#	105#	120#	135#	150#
4#4oz.	8#8oz.	12#12oz.	17#	34#	51#	68#	85#	102#	119#	136#	153#	170#
4#8oz.	9#	13#8oz.	18#	36#	54#	72#	90#	108#	126#	144#	162#	180#
5#	10#	15#	20#	40#	60#	80#	100#	120#	140#	160#	180#	200#
5#12oz.	11#8oz.	17#4oz.	23#	46#	69#	92#	115#	138#	161#	184#	207#	230#
6#4oz.	12#8oz.	18#12oz.	25#	50#	75#	100#	125#	150#	175#	200#	225#	250#

*The material in these charts was developed by the New York Department of Mental Health, and furnished through the courtesy of Mrs. Katherine Flack, Director of Nutrition Service to Koch Refrigerators, Inc.

INDEX

Addend, 27
Addition, 27–31
  decimals, 61–62
  denominate numbers, 73
  fractions, 52–55
    like fractions, 52
    unlike fractions, 53–55
    with whole numbers, 66
  pound/ounces, 167
  table for, 11
A la carte menu, 172
Apprentices, wage scales, 117
Assets, 192

Balance sheet, 192–194
  net worth, finding, 193
  terms related to, 193
Book value, 200
Branching, factorization, 37
Break-even analysis, 154–156
  fixed expenses, 154
  formula for, 155–156
  new businesses, 155
  and seating turnover, 158
  time period for, 155
  variable expenses, 154
Business forms
  invoices, 184
    discount terms, 186
  purchase orders, 183
  purchase specifications, 184, 185
  requisitions, 181–183
  See also Daily reports; Financial statements.
Business ownership, 219–222
  co-operative, 220–221
  corporation, 219–220, 221
  partnership, 219, 221
  single proprietorship, 219, 221

Cash receipts report, 179–180
Change, making change, 45–46
Check, average check, finding, 50–51
Checks. *See* Guest checks
Collective bargaining, 116
Common denominator, fractions, finding, 55
Common fraction, 52
Comparative Summary, 152

Complex fraction, 52
Composite numbers, 37
Consumer price index (CPI), 119–121
  and market basket, 119–120
  use of, 120–121
Container sizes, 256–258
Conversion tables, 236–237
  fractional equivalents, 234
Cook's Counter Report, 148–149
Cook's Production Report, 147

Daily reports
  cook's, 147–153
    Comparative Summary, 152
    Counter Report, 148–149
    with double check, 150
    Production Report, 147
    Summary of all Departments, 152
Days, expressed as decimals, 242
Decimal fractions, 60
Decimal point, 25
Decimals, 60–65
  addition, 61–62
  days expressed as, 242
  division, 62–65
  equivalents of fractions, 238
  fractions, changing to decimals, 60–61
  multiplication, 62
  subtraction, 61–62
Denominate numbers
  addition, 73
  division, 74–75
  multiplication, 74
  subtraction, 73–74
Denominator, 52
Depreciation, 199–200
  book value, 200
  calculation of, 200
  government regulations, 199
  salvage value, 200
  straight-line method, 200
Difference/remainder, 41
Discounts
  cash discount, 186
  quantity discount, 186
Dish-ups, standard, 134
Dividend, 47

Division, 47–51
  average check, finding, 50–51
  decimals, 62–65
  denominate numbers, 74–75
  division by subtraction, 48–50
  factoring and, 40
  fractions, 70–72
Divisor, 47

Egg specifications, table for, 232
Employee taxes
  payroll records, 208–211
  social security tax, 212–213
Estimated tax payments, 204

Factorization, 37–40
Financial statements
  balance sheet, 192–194
  and depreciation, 199–200
  income statement, 194–199
Fractions, 52–59
  addition, 52–55
    like fractions, 52
    unlike fractions, 53–55
    with whole numbers, 66
  changing to decimals, 60–61
  common denominator, finding, 55
  decimal equivalents table, 15, 238
  division, 70–72
  mixed numbers, changing, 58–59
  multiplication, 68–70
  reducing/raising, to higher terms, 55–56
  subtraction, 56–58
    like fractions, 56
    unlike fractions, 57–58
    with whole numbers, 66–68
Fringe benefits, 128–129
Frozen food capacity, 258

Gourmand, 164
Gourmet, 164
Griddle items, cooking guide, 259
Guest checks, 169–178
  cover charge, 173
  handwritten, 172
  minimum charge, 173

partly preprinted, 169
preprinted, 169
sales tax, determining, 170
tip, 173
writing the order, 173

Hiring, criteria for, 225

Improper fraction, 52
Income statement
   percentage of sales, calculation of, 197
   "set of books," information from, 195
Inflation, 119, 121–122
   causes of, 121–122
   consumer price index (CPI), 119–121
   escalator clause, 121
   government intervention, 122
   in USA, historical view, 121
Inventory
   first in, first out, 191
   par stock, 190–191
   perpetual inventory, 190
   physical inventory, 190

Labor costs, 125–127
   meals in, 126
Liabilities, 192
Loans, 202–203
   interest, change in amount of, 202–203
   repayment, 202
Local taxes, 214–216
Lowest common denominator, 38–39
   equivalent fractions table, 15
   finding, 55
   number line for, 54
Lowest common multiple, 39

Mark-up, pricing the menu, 160–162
Mathematical symbols, listing of, 9–10
Measurements
   abbreviations and equivalents, 243
   common foods, approximate weights/measures of, 240–241
   container sizes, 256–258
   tables of measure, 231
   tablespoons per ounce, listing of common foods, 255
   weight conversion charts, 260–261
   weight/volume
      listing of, 163
      listing of common foods, 244–246
   See also Denominate numbers.

Menus
   a la carte menu, 172
   table d'hote menu, 172
Minuend, 41
Mixed numbers, 52
   changing, for addition/subtraction, 58–59
Multiplicand, 32, 36
Multiplication, 32–40
   decimals, 62
   denominate numbers, 74
   factorization, 37–40
   fractions, 68–70
   table for, 12–13
Multiplier, 32, 36

Net worth, finding, 193
Numbers/numerals, 23–26
   number lines, 26
   place value, 23–25
Numerator, 52

Ounces
   meanings of, 163
   ounces/eighths of ounces, as decimal parts of pound (table), 239

Payroll records, 208–211
Penalties, taxes, 205
Percentage, 80–81
   and ratio/proportion, 79–81
   yield percentages, 138–139
Percentage of sales, calculation of, 197
Perpetual inventory, 190
Physical inventory, 190
Place value, 23–25
Portions
   container portion/conversion chart, 235
   portion control data, 251–255
   size/utensils needed, table, 247–250
   standard, 133–134
      measuring portions, ladels/scoops/spoons, 233
Pricing the menu, 160–162
Prime factors, 37, 38
Prime numbers, table of, 16
Procedures, standard, 134
Product, 32, 36
Proper fraction, 52
Property taxes, 215
Proportion, 76, 77–80

Quotient, 47

Ratio, 76
   and proportion, 76, 77–80
Rational numbers, 52, 60, 76

Recipes
   converting, 135–137
      conversion tables, 236–237
      fractional equivalents, 234
   costing of, 143–146
   standard, 133
Reciprocal of number, 52, 70
Regrouping. *See* Subtraction

Sales tax, determining, 170
Salvage value, 200
Sanitation
   bacteria, growth of, 113–114
   guidelines for, 111–114
Scale drawing, plans for inn, 223–224
Seating turnover, 157–159
   and break-even analysis, 158
   new businesses, 157–158
"Set of books," information from, 195
Social security tax, employee taxes, 212–213
Standards in food service, 133–134
   dish-ups, 134
   portions, 133–134
   procedures, 134
   recipes, 133
   yields, 133
   See also Portions; Recipes; Yields.
State income taxes, 207–208
Storekeeper, 190–191
   duties of, 190
Storeroom, inventory, 190–191
Straight-line method, 200
Subtraction, 41–44
   decimals, 61–62
   denominate numbers, 73–74
   fractions, 56–58
      like fractions, 56
      unlike fractions, 57–58
      with whole numbers, 66–68
   regrouping
      borrowing, 42–43
      subtraction with zeros, 43–44
   table for, 14
Subtrahend, 41
Sum, 27
Summary of all Departments, 152

Table d'hote menu, 172
Taxes, 204–218
   employee taxes
      payroll records, 208–211
      social security tax, 212–213
   estimated tax payments, 204
   filing requirements, 205
   on free meals/lodging, 206
   local taxes, 214–216

penalties, 205
property taxes, 215
social security numbers and, 207
state income taxes, 207–208
taxable income, 205
tax preparation, help in, 207
on tips, 206
Terms of a fraction, 52
Tipping
   standard custom for, 173
   and taxes, 206
   wage scales and, 116

Wage scales, 116–118
   apprentices, 117
   collective bargaining, 116
   nontipped/tipped employees, 116
   work day, determination of, 116–117
Whole numbers, 25
Wine, 164–166
   bottle sizes, 165
   glass sizes, 165
   house wine, 165
   varieties of, 164
   wine list, 164
Work day, determination of, 116–117
Workmen's compensation insurance, 129–130

Yields
   as purchased compared to edible portion, 138–139
   standard, 133
   yield percentages, 138–139